AF251335

Asymptotic Methods in
Nonlinear Wave
Phenomena

In honor of the 65th birthday of
Antonio Greco

Asymptotic Methods in Nonlinear Wave Phenomena

In honor of the 65th birthday of
Antonio Greco

Palermo, Italy 5 – 7 June 2006

Editors

T Ruggeri
University of Bologna, Italy

M Sammartino
University of Palermo, Italy

World Scientific

NEW JERSEY · LONDON · SINGAPORE · BEIJING · SHANGHAI · HONG KONG · TAIPEI · CHENNAI

Published by

World Scientific Publishing Co. Pte. Ltd.

5 Toh Tuck Link, Singapore 596224

USA office: 27 Warren Street, Suite 401-402, Hackensack, NJ 07601

UK office: 57 Shelton Street, Covent Garden, London WC2H 9HE

British Library Cataloguing-in-Publication Data
A catalogue record for this book is available from the British Library.

ASYMPTOTIC METHODS IN NONLINEAR WAVE PHENOMENA
In Honor of Antonio Greco on his 65th Birthday

ISBN-13 978-981-270-782-6
ISBN-10 981-270-782-4

Printed in Singapore by World Scientific Printers (S) Pte Ltd

PREFACE

This volume is dedicated to Antonio Greco's influence and achievements, on his 65th birthday. It is an outgrowth from a conference on *Asymptotic Methods in Nonlinear Wave Phenomena* which was held in his honor in Mondello (Palermo) 5-7 June 2006.

Antonio has made significant contributions to the understanding of nonlinear propagation phenomena and to symmetry group methods applied to PDEs arising in mathematical physics. Here we will just mention his research on completely exceptional hyperbolic systems and Von Karman's fluids; asymptotic waves; the interaction between gravitational and acoustic waves in relativistic fluid dynamics and the completely integrability of the generalized Kadomtsev-Petviashvili equation through the use of the Painlevé test and of the Lax pair.

His merit also includes the gathering in Palermo of a group of young and enthusiastic researchers sharing his vision and scientific interests. We believe that, under his leadership, this group will continue to grow and to accomplish even more ambitious goals.

Antonio Greco has held several administrative responsibilities. He has served as a Director of a CNR laboratory and is currently the Director of the Math Department and also the Principal Investigator of the unit of Palermo in the PRIN funding project. He has organized several successful international conferences and taught twice, in 1982 and in 1986, at the "Scuola Estiva di Fisica Matematica" in Ravello, the last being jointly taught with our unforgetable colleague and friend, Andrea Donato.

It is also a pleasure to mention Antonio's unique qualities which have made him a special man. He has a bright personality, a generous nature and it is always a pleasure being in his company.

We wish him all the best and to keep up the good work.

He will definitely remain as brotherly friend for the rest of our lives.

Tommaso Ruggeri and Marco Sammartino

January 2007

Professor Antonio Greco
Università degli Studi di Palermo

SCIENTIFIC COMMITTEE

Chairman: **T. Ruggeri** (Bologna)
Y. Choquet-Bruhat (Parigi), **S. Rionero** (Napoli)
M. Sammartino (Palermo)

ORGANIZING COMMITTEE

Chairman: **M. Sammartino** (Palermo)
G. Gambino (Palermo), **M. C. Lombardo** (Palermo),
V. Sciacca (Palermo), **L. Seta** (Palermo)

SUPPORTED BY

- Università degli Studi di Palermo
- Gruppo Nazionale per la Fisica Matematica (G.N.F.M.) dell'INdAM
- MIUR PRIN 2003-05 Nonlinear Mathematical Problems of Wave Propagation and Stability in Models of Continuous Media

CONTENTS

THE SEMICONDUCTOR STEADY BOLTZMANN EQUATION: A VARIATIONAL FORMULATION WITH AN APPLICATION TO MOBILITY

A.M. ANILE

Dipartimento di Matematica ed Informatica,
Università degli Studi di Catania, Italy,
E-mail: anile@dmi.unict.it

G. ALÌ

Istituto per le Applicazioni del Calcolo "M. Picone", sez. di Napoli,
Consiglio Nazionale delle Ricerche,
via Pietro Castellino 111, 80131 Napoli, Italy,
and INFN-Gruppo c. Cosenza, Italy,
E-mail: g.ali@iac.cnr.it

G. MASCALI

Dipartimento di Matematica, Università della Calabria,
ponte Bucci, cubo 30 B, 87036 Arcavacata di Rende, Italy
and INFN-Gruppo c. Cosenza, Italy,
E-mail: g.mascali@unical.it

We consider a variational formulation of the steady Boltzmann equation for semiconductors. We apply this formulation to the calculation of an approximate expression of the electron mobility, valid for electric fields up to 1-1.5 $\frac{V}{\mu m}$.

Keywords: Semiclassical Boltzmann equation; Semiconductors; Variational formulation; Mobility.

1. Introduction

The semiclassical Boltzmann (BE) equation is the reference model for charge transport in semiconductors.[1-3] In fact, not only the results of simulations by other models are usually compared with BE simulations, but most of the existing models in microelectronics have been or can be derived from BE.[1,4]

In the past, calculus of variations has been extensively applied to transport theory, mainly to obtain estimates of physical quantities or to derive approximate methods for transport problems.[5,6] These variational approach has been particularly fruitful in the field of kinetic models of plasmas, with application in nuclear engineering and reactor analysis.[7]

Despite of the great interest of the Boltzmann equation in semiconductor applications, so far variational methods have not been applied to this model, with few exceptions. Isolated attempts in this direction can be found in Ref. 8, where a variational formulation of the BTE is introduced to evaluate an approximate expression of the mobility for a bulk semiconductor, and, more recently, in Ref. 9, where the mathematical setting of the variational formulation has been clarified.

In this paper, after a brief review of the variational formulation of the steady-state BE, we show how this formulation can be used for studying the behavior of the electron mobility in bulk silicon as a function of the electron average energy. The formula is valid at low energies, more general approximations are under investigation and will be presented as soon as possible.

2. Boltzmann transport equation

At kinetic level a family of carriers in an infinitely extended semiconductor is described by a distribution function

$$f(\mathbf{x}, \mathbf{k}, t), \text{ with } \mathbf{x} \in \mathbb{R}^3, \mathbf{k} \in \mathcal{B} \subset \mathbb{R}^3, \text{ and } t \in \mathbb{R}^+.$$

Here, $\mathcal{B}$ is a bounded set, the first Brillouin region of the inverse lattice of the semiconductor crystal. The carrier energy is expressed in terms of the wave vector $\mathbf{k}$ by means of a dispersion relation $\mathcal{E}(\mathbf{k})$, it is determined by the band structure of the semiconductor, and satisfies the following symmetry property

$$\mathcal{E}(-\mathbf{k}) = \mathcal{E}(\mathbf{k}), \tag{1}$$

due to the time reversibility of the Schrödinger equation.[10] Moreover, we assume that:

$$\text{the isoenergetic surfaces, } \mathcal{E}(\mathbf{k}) = \text{constant, are closed and bounded.} \tag{2}$$

The dispersion relation defines the group velocity of the carriers (Bloch electrons or holes),

$$\mathbf{v}(\mathbf{k}) = \frac{1}{\hbar} \nabla_{\mathbf{k}} \mathcal{E}(\mathbf{k}),$$

which in the semiclassical approximation is identified with the carrier velocity.

The time evolution of the distribution function f is determined by the semiclassical Boltzmann-Poisson system

$$\frac{\partial f}{\partial t} + \mathbf{v} \cdot \nabla_{\mathbf{x}} f + \frac{q_c}{\hbar} \mathbf{E} \cdot \nabla_{\mathbf{k}} f = C[f], \tag{3}$$

$$-\nabla_{\mathbf{x}} \cdot (\epsilon_s \nabla_{\mathbf{x}} \phi) = Q_{\mathrm{bi}} + q_c \int_{\mathcal{B}} f d\mathbf{k}, \tag{4}$$

where $\mathbf{E}$ is the electric field, $C[f]$ the collision operator, ϕ the electric potential, q_c the carrier's charge, Q_{bi} the built-in charge, related to the dopant concentration.

The electric field is related to the electric potential by the usual relation

$$\mathbf{E} = -\nabla_{\mathbf{x}} \phi. \tag{5}$$

The collision operator takes into account various scattering mechanisms, in particular carrier-phonon scattering and carrier-impurity scattering. In the non-degenerate case, it can be written in the general form:

$$C[f] = \int_{\mathcal{B}} \left[P(\mathbf{k}', \mathbf{k}) f(\mathbf{x}, \mathbf{k}', t) - P(\mathbf{k}, \mathbf{k}') f(\mathbf{x}, \mathbf{k}, t) \right] d\mathbf{k}', \tag{6}$$

where the kernel $P(\mathbf{k}', \mathbf{k})$, which is the transition probability per unit time from a state $\mathbf{k}$ to a state $\mathbf{k}'$, satisfies the detailed balance principle

$$P(\mathbf{k}', \mathbf{k}) \exp\left[-\mathcal{E}(\mathbf{k}')/k_B T_L\right] = P(\mathbf{k}, \mathbf{k}') \exp\left[-\mathcal{E}(\mathbf{k})/k_B T_L\right], \tag{7}$$

and the symmetry condition

$$P(-\mathbf{k}', -\mathbf{k}) = P(\mathbf{k}', \mathbf{k}). \tag{8}$$

Equations (3), (4) are defined for $\mathbf{x} \in \mathbb{R}^3$. The same equations can be considered in a bounded subset $\Omega \subset \mathbb{R}^3$. In this case, appropriate boundary conditions are needed. In general, the boundary can be split as $\partial \Omega = \Gamma_D \cup \Gamma_N$, where Γ_D comprises the Ohmic boundaries and Γ_N the insulating boundaries, with $\Gamma_D \cap \Gamma_N = \emptyset$. We denote by $\boldsymbol{\nu}(\mathbf{x})$ the external normal to the boundary at $\mathbf{x} \in \partial \Omega$, and introduce the following notation:

$$\mathcal{B}^{\pm}(\mathbf{x}) = \left\{ \mathbf{k} \in \mathcal{B} \mid \pm \mathbf{v}(\mathbf{k}) \cdot \boldsymbol{\nu}(\mathbf{x}) > 0 \right\},$$
$$\Sigma_{D,N}^{\pm} = \left\{ (\mathbf{x}, \mathbf{k}) \in \Gamma_{D,N} \times \mathcal{B} \mid \mathbf{k} \in \mathcal{B}^{\pm}(\mathbf{x}) \right\}, \quad \Sigma^{\pm} = \Sigma_N^{\pm} \cup \Sigma_D^{\pm},$$
$$f^{\pm} = f\big|_{\Sigma^{\pm} \times \mathbb{R}^+}.$$

The boundary conditions are assigned on the incoming carriers, that is, on the function f^-. For the Ohmic boundaries, one usually assigns Dirichlet conditions of the form:

$$f^-(\mathbf{x}, \mathbf{k}, t) = \bar{f}(\mathbf{x}, \mathbf{k}, t), \quad (\mathbf{x}, \mathbf{k}) \in \Sigma_D^-, \, t > 0, \tag{9}$$

4

where $\bar{f}$ is a given function, usually the lattice temperature Maxwellian. For insulating boundaries, it is appropriate to assign Neumann conditions of the form:

$$f^-(\mathbf{x},\mathbf{k},t) = \tilde{\mathcal{K}}_N[f^+](\mathbf{x},\mathbf{k},t), \quad (\mathbf{x},\mathbf{k}) \in \Sigma_N^-, \, t > 0, \tag{10}$$

where

$$\tilde{\mathcal{K}}_N[f^+](\mathbf{x},\mathbf{k},t) = \int_{\mathcal{B}^+(\mathbf{x})} \tilde{R}(\mathbf{k}',\mathbf{k}) f^+(\mathbf{x},\mathbf{k}',t)\,d\mathbf{k}',$$

and the kernel $\tilde{R}(\mathbf{k}',\mathbf{k})$ satisfies the reciprocity property

$$\tilde{R}(\mathbf{k}',\mathbf{k})|\mathbf{v}(\mathbf{k})\cdot\boldsymbol{\nu}(\mathbf{x})| = \tilde{R}(-\mathbf{k},-\mathbf{k}')|\mathbf{v}(\mathbf{k}')\cdot\boldsymbol{\nu}(\mathbf{x})|, \quad \forall\, \mathbf{k} \in \mathcal{B}^-(\mathbf{x}). \tag{11}$$

The boundary conditions can be rewritten more concisely, as

$$f^- = 1_{\Gamma_N}\tilde{\mathcal{K}}_N[f^+] + 1_{\Gamma_D}\bar{f}, \quad \text{in } \Sigma^- \times \mathbb{R}^+. \tag{12}$$

where $1_{\Gamma_{N,D}}$ represent the characteristic functions of the sets $\Gamma_{N,D}$. Further details can be found in Ref. 11.

3. Variational formulation

In this section we derive a variational formulation of the steady Boltzmann equation for semiconductors when $\phi(\mathbf{x})$ is a time independent externally applied potential.

To start with, we introduce a weighted distribution function, h, defined by

$$f(\mathbf{x},\mathbf{k}) = M_\phi(\mathcal{E}(\mathbf{k}))\, h(\mathbf{x},\mathbf{k}),$$

where $M_\phi(\mathcal{E})$ is the Maxwell equilibrium distribution,

$$M_\phi(\mathcal{E}) = \exp\left[-\frac{\mathcal{E}+q_c\phi}{k_B T_L}\right].$$

Also, we introduce the streaming operator

$$\mathcal{D} = \mathbf{v}\cdot\nabla_{\mathbf{x}} + \frac{q_c}{\hbar}\mathbf{E}\cdot\nabla_{\mathbf{k}}.$$

By using the detailed balance principle (7), and the identity $\mathcal{D}M_\phi(\mathcal{E}) = 0$, it is possible to see that the weighted distribution function $h(\mathbf{x},\mathbf{k})$ satisfies the modified Boltzmann equation

$$\mathcal{D}h = \mathcal{C}[h], \tag{13}$$

where

$$C[h] = \int_{\mathcal{B}} P(\mathbf{k}, \mathbf{k}') \left[h(\mathbf{x}, \mathbf{k}') - h(\mathbf{x}, \mathbf{k})\right] d\mathbf{k}'.$$

As a result of this change of coordinate, the new collision operator $\mathcal{C}$ is symmetric with respect to the weighted scalar product

$$((h, g)) = \int_{\Omega} \int_{\mathcal{B}} h(\mathbf{x}, \mathbf{k}) \, M_{\phi(\mathbf{x})}(\mathcal{E}(\mathbf{k})) \, g(\mathbf{x}, \mathbf{k}) \, d\mathbf{k} \, d\mathbf{x}, \tag{14}$$

that is,

$$((h, \mathcal{C}[g])) = ((\mathcal{C}[h], g)). \tag{15}$$

Unfortunately, the streaming operator $\mathcal{D}$ is not symmetric with respect to the same scalar product (14). Anyway, by using a well known trick,[5] $\mathcal{D}$ can be symmetrized by means of the parity operator $\mathcal{P}$, defined by

$$\mathcal{P}h(\mathbf{x}, \mathbf{k}) = h(\mathbf{x}, -\mathbf{k}). \tag{16}$$

In fact, by integration by parts, it is possible to show that $\mathcal{P}\mathcal{D}$ satisfies the symmetry condition

$$((\mathcal{P}\mathcal{D}h, g)) + ((\mathcal{P}h, g))_{\Sigma^-} = ((h, \mathcal{P}\mathcal{D}g)) + ((h, \mathcal{P}g))_{\Sigma^-}, \tag{17}$$

where

$$((h, g))_{\Sigma^-} = \int_{\partial\Omega} \int_{\mathcal{B}^-(\mathbf{x})} h(\mathbf{x}, \mathbf{k}) \, M_{\phi(\mathbf{x})}(\mathcal{E}(\mathbf{k})) \, g(\mathbf{x}, \mathbf{k}) |\mathbf{v}(\mathbf{k}) \cdot \boldsymbol{\nu}(\mathbf{x})| d\mathbf{k} \, d\sigma_x,$$
$$\tag{18}$$

with $d\sigma_x$ surface element on the boundary.

Moreover, the symmetry of $\mathcal{C}$ is not destroyed by the parity operator. In fact, the symmetry condition (8) implies that the operator $\mathcal{P}\mathcal{C}$ is also symmetric,

$$((h, \mathcal{P}\mathcal{C}[g])) = ((\mathcal{P}\mathcal{C}[h], g)). \tag{19}$$

Then, it is convenient to replace the modified Boltzmann equation (13) with the equivalent, symmetric equation

$$\mathcal{P}\mathcal{D}h = \mathcal{P}\mathcal{C}[h]. \tag{20}$$

This equation is the starting point for a variational formulation of the steady-state Boltzmann equation.

It is tempting to introduce the functional

$$\tilde{J}(h) = ((h, \mathcal{P}\mathcal{D}h - \mathcal{P}\mathcal{C}[h])). \tag{21}$$

6

Then, by using the symmetry conditions, it is possible to prove that

$$\delta \tilde{J}(h) = 2\left((\delta h, \mathcal{P}\mathcal{D}h - \mathcal{P}\mathcal{C}[h])\right) - ((\mathcal{P}\delta h, h))_{\Sigma^-} + ((\delta h, \mathcal{P}h))_{\Sigma^-} . \qquad (22)$$

This shows that the solutions of the modified Boltzmann equation (20) are not stationary points of $\tilde{J}$. Thus, we need to deal with boundary conditions.

The boundary conditions relative to the insulating boundaries can be written

$$h(\mathbf{x}, \mathbf{k}) = \mathcal{K}_N[h](\mathbf{x}, \mathbf{k}) := \frac{\tilde{\mathcal{K}}_N[M_\phi(\mathcal{E})h](\mathbf{x}, \mathbf{k})}{M_{\phi(\mathbf{x})}(\mathcal{E}(\mathbf{k}))}, \quad (\mathbf{x}, \mathbf{k}) \in \Sigma_N^- .$$

Therefore, taking $\bar{f}(\mathbf{x}, \mathbf{k}) = \bar{\psi}(\mathbf{x})M_\phi(\mathcal{E}(\mathbf{k}))$, the boundary conditions (12) become

$$h = 1_{\Gamma_N}\mathcal{K}_N[h] + 1_{\Gamma_D}\bar{\psi}, \quad \text{in } \Sigma^- . \qquad (23)$$

By using the reciprocity (11), it is possible to see that $\mathcal{K}_N$ satisfies the symmetry property

$$((\mathcal{P}g, \mathcal{K}_N[h]))_{\Sigma^-} = ((\mathcal{K}_N[g], \mathcal{P}h))_{\Sigma^-} .$$

In conclusion, introducing the functional

$$J(h) = \tilde{J}(h) + \left(((\mathcal{P}h, h - 1_{\Gamma_N}\mathcal{K}_N[h] - 1_{\Gamma_D} 2\bar{\psi}))\right)_{\Sigma^-} , \qquad (24)$$

it is possible to show that

$$\delta J(h) = 2\left((\delta h, \mathcal{P}\mathcal{D}h - \mathcal{P}\mathcal{C}[h])\right) + 2\left((\mathcal{P}\delta h, h - 1_{\Gamma_N}\mathcal{K}_N[h] - 1_{\Gamma_D}\bar{\psi}))\right)_{\Sigma^-} . \qquad (25)$$

As a result of (25), the following theorem holds:

Theorem 1. *The weighted distribution function h is solution of the modified Boltzmann equation (20), with boundary conditions (23), if and only if it is a stationary point of the functional J, that is,*

$$\delta J(h) = 0. \qquad (26)$$

4. Mobility in bulk Silicon

In this section, we want to use the variational formulation (26) in order to find an approximate expression for the electron mobility in bulk silicon, that is a uniformly doped piece of semiconductor. Therefore in the following we take $q_c = -q$, and $Q_{\text{bi}} = qN_D$, with q the elementary charge and N_D the donor concentration. The hole mobility can be found in a similar way, by taking $q_c = q$, and $Q_{\text{bi}} = -qN_A$, with N_A acceptor concentration.

First of all, we can consider Ω extended to the whole space so that the boundary conditions can be neglected. We take into account the interactions of the electrons in the six equivalent valleys of the conduction band with phonons and impurities.[12] Some of the interactions with phonons leave the electrons in the same valley as they are before the collision (intravalley transitions), while others of them can drive the electrons into a different valley (intervalley transitions) according to suitable selection rules. The transition rate of the collision operator, $P(\mathbf{k}, \mathbf{k}')$, is therefore the sum of

- the transition rate for the electron-acoustical phonon intravalley scattering:

$$P_{ac}(\mathbf{k}, \mathbf{k}') = \mathcal{K}_{ac}\, \delta(\mathcal{E}' - \mathcal{E}),$$

  written in the elastic approximation, which is valid when the thermal energy is much greater that of the phonon. Here, $\mathcal{K}_{ac}$ is the acoustical intravalley scattering kernel coefficient and δ the Dirac function;

- the six intervalley scattering transition rates:

$$P_\alpha(\mathbf{k}, \mathbf{k}') = \mathcal{K}_\alpha \left[n_\alpha\, \delta(\mathcal{E}' - \mathcal{E} - \hbar\omega_\alpha) + (n_\alpha + 1)\delta(\mathcal{E}' - \mathcal{E} + \hbar\omega_\alpha) \right],$$

  where α runs over the three g_1, g_2, g_3 and the three f_1, f_2, f_3 intervalley scatterings.[12] The $\mathcal{K}_\alpha$'s are the optical or acoustical intervalley scattering kernel coefficients and

$$n_\alpha = \frac{1}{exp\left(\frac{\hbar\omega_\alpha}{k_B T_L}\right) - 1}$$

  is the occupation number of phonons with frequency ω_α;

- the impurity scattering transition rate:

$$P_{imp}(\mathbf{k}, \mathbf{k}') = \frac{\mathcal{K}_{imp}}{\left[|\mathbf{k} - \mathbf{k}'|^2 + \beta^2\right]^2}\, \delta(\mathcal{E}' - \mathcal{E}),$$

  with $\mathcal{K}_{imp}$ scattering kernel coefficient and β inverse Debye length.

Now, we assume that f has the form

$$f(\mathbf{x}, \mathbf{k}) = f^{(0)}(\mathbf{k})\left[1 + \Phi(\mathbf{k})\right], \tag{27}$$

with

$$f^{(0)}(\mathbf{k}) = \frac{n}{N}\exp\left(-\lambda_W\, \mathcal{E}(\mathbf{k})\right), \tag{28}$$

and

$$\Phi(\mathbf{k}) = -\frac{3q}{|\mathbf{v}(\mathbf{k})|^2}\, \mu\, \mathbf{v}(\mathbf{k}) \cdot \mathbf{E}. \tag{29}$$

8

Here, $n = \int f d\mathbf{k}$ is the carrier density, N the normalization factor, μ the mobility and λ_W is related to the mean energy $W = \frac{\int \mathcal{E} f d\mathbf{k}}{\int f d\mathbf{k}}$ through the relation

$$W = \frac{1}{N} \int_{\mathcal{B}} \mathcal{E} \exp(-\lambda_W \mathcal{E}) \, d\mathbf{k}.$$

We also assume that the dispersion relation is given by the Kane approximation

$$\mathcal{E}(1 + \alpha \mathcal{E}) = \frac{\hbar^2 |\mathbf{k}|^2}{2 m^*},$$

with α non parabolicity factor.

Using ansatz (27), (28) and (29) in the variational formulation amounts to choosing

$$\tilde{h} = \frac{n}{N} \exp\left(\frac{\mathcal{E} - q\mathbf{E}\cdot\mathbf{x}}{k_B T_L} - \lambda_W \mathcal{E}\right) \left[1 - \frac{3q}{|\mathbf{v}|^2} \mu \, \mathbf{v} \cdot \mathbf{E}\right].$$

Taking variations with respect to μ, we obtain the following expression for the mobility

$$\mu = -\frac{((\psi_1, \mathcal{P}(\mathcal{D} - \mathcal{C})[\psi_0]))}{((\psi_1, \mathcal{P}(\mathcal{D} - \mathcal{C})[\psi_1]))}, \tag{30}$$

where

$$\psi_0 = \frac{n}{N} \exp\left(\frac{\mathcal{E} - q\mathbf{E}\cdot\mathbf{x}}{k_B T_L} - \lambda_W \mathcal{E}\right),$$

and

$$\psi_1 = -\frac{3n}{N|\mathbf{v}|^2} \exp\left(\frac{\mathcal{E} - q\mathbf{E}\cdot\mathbf{x}}{k_B T_L} - \lambda_W \mathcal{E}\right) \mathbf{v} \cdot \mathbf{E}.$$

The formula (30) expresses μ as a function of W, T_L and of the doping concentration.

After calculations, we find

$$\mu = \frac{c_1 \, d(\lambda_W)}{c_2 \left[d_{ac}(\lambda_W) + \sum_\alpha d_\alpha(\lambda_W)\right] + c_3 \, d_{imp}(\lambda_W)},$$

where

$$c_1 = \frac{\sqrt{2}q}{3\hbar}, \quad c_2 = \frac{4\pi m^{*\frac{5}{2}}}{\hbar^4}, \quad c_3 = \frac{\pi m^{*\frac{1}{2}}}{8},$$

and

$$d = \lambda_W \int_0^\infty e^{-\mathcal{E}(2\lambda_W - \lambda_W^0)}(1 + 2\alpha\mathcal{E})\sqrt{\mathcal{E}(1 + \alpha\mathcal{E})}\, d\mathcal{E},$$

$$d_{ac} = \kappa_{ac} \int_0^\infty e^{-\mathcal{E}(2\lambda_W - \lambda_W^0)}(1 + 2\alpha\mathcal{E})^4\, d\mathcal{E},$$

$$d_\alpha = \kappa_\alpha \int_0^\infty e^{-\mathcal{E}(2\lambda_W - \lambda_W^0)} \left\{ n_\alpha(1 + 2\alpha\mathcal{E})^3(1 + 2\alpha\mathcal{E}_\alpha^+)\sqrt{\frac{\mathcal{E}_\alpha^+(1 + \alpha\mathcal{E}_\alpha^+)}{\mathcal{E}(1 + \alpha\mathcal{E})}} \right.$$

$$\left. + (n_\alpha + 1)(1 + 2\alpha\mathcal{E}_\alpha^+)^3(1 + 2\alpha\mathcal{E})\sqrt{\frac{\mathcal{E}(1 + \alpha\mathcal{E})}{\mathcal{E}_\alpha^+(1 + \alpha\mathcal{E}_\alpha^+)}} \right\} d\mathcal{E},$$

$$d_{imp} = \kappa_{im} \int_\mathcal{B} e^{-\mathcal{E}(2\lambda_W - \lambda_W^0)}(1 + 2\alpha\mathcal{E})^4 \frac{[1 + g(\mathcal{E})]\ln[1 + g(\mathcal{E})] - g(\mathcal{E})}{[1 + g(\mathcal{E})]\mathcal{E}^2(1 + \alpha\mathcal{E})^2}\, d\mathcal{E},$$

with $\lambda_W^0 = \frac{1}{k_B T_L}$, $\mathcal{E}_\alpha^+ = \mathcal{E} + \hbar\omega_\alpha$, $g(\mathcal{E}) = 8a\mathcal{E}(1 + \alpha\mathcal{E})$, $a = \frac{m^*}{\hbar^2\beta^2}$.

In Fig.1, we show the behavior of mobility as a function of the average energy, neglecting the contribution from the scattering with impurities.

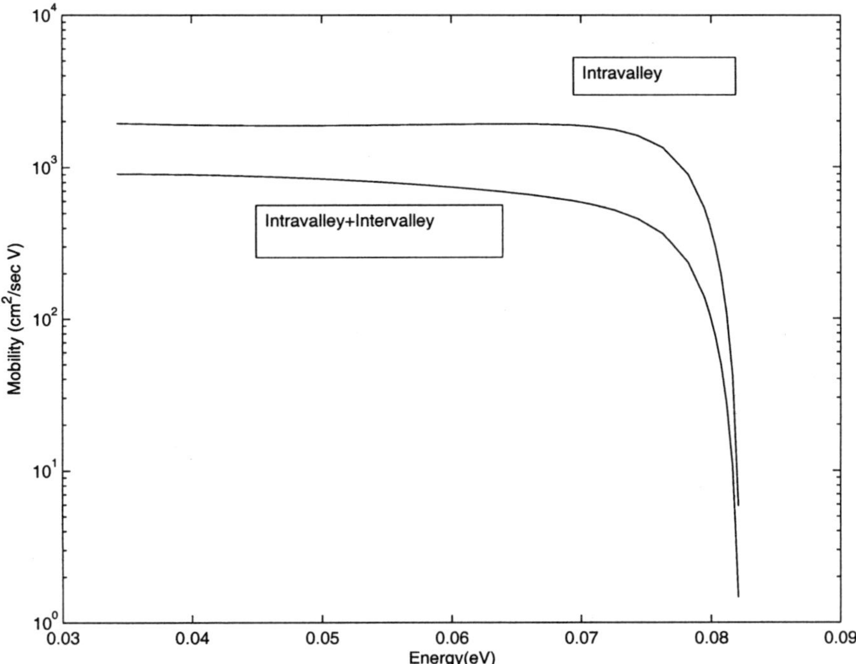

Fig. 1.   Mobility vs energy (bulk Silicon).

5. Conclusions

As we have anticipated in the introduction, the main aim of this paper is to attract attention on variational methods in semiclassical transport models. We believe that these methods could be used for developing new numerical approximations of the semiclassical Boltzmann equation for semiconductors. In this respect, the main mathematical problem is the lack of any coercivity property of the functional J.

As regards the application which we have considered in greater extent in this paper, the main drawback of our implementation is the lack of validity for $\lambda_W \geq \frac{1}{2 k_B T_L}$, due to presence of the weight function in the functional. This means that the ansatz used for our approximation is good only for low electric fields, up to 1-1.5 $\frac{V}{\mu m}$. However this problem can be tackled by making a more suitable ansatz. For example, one could take the distribution function in the form suggested by the Maximum Entropy Principle (MEP)[13]

$$f(\mathbf{k}) = \exp\left(-\sum_{n=0}^{N} \lambda_{W_n} \mathcal{E}^n - \mathbf{v} \cdot \sum_{m=0}^{M} \lambda_{\mathbf{S}_m} \mathcal{E}^m\right)$$

where $N \geq 2$ and M can be suitably chosen and the λ_W's and $\lambda_\mathbf{S}$'s, Lagrange multipliers, depend on the correspondent moments of the MEP distribution

$$n = \int f d\mathbf{k}, \qquad W_n = \frac{\int \mathcal{E}^n f d\mathbf{k}}{\int f d\mathbf{k}}, \quad n = 1, \ldots, N,$$

$$\mathbf{S}_m = \frac{\int \mathcal{E}^m \mathbf{v} f d\mathbf{k}}{\int f d\mathbf{k}}, \quad m = 0, \ldots, M,$$

W_1 is the energy, $\mathbf{S}_0$ and $\mathbf{S}_1$ the velocity and energy flux respectively.

With this ansatz for the distribution function, the problem is reduced to finding the stationary points of a function of N+3M-1 variables: λ_{W_n}, $\lambda_{\mathbf{S}_m}$, $n = 1, \ldots, N$, $m = 0, \ldots, M$, λ_{W_0} being determined in terms of the remaining λ_W's by the condition $n = N_D$.

The numerical solution of this problem, for various values of N and M, is still in progress and will be presented in a forthcoming paper.[11]

Dedication

$\tau\upsilon\gamma\chi\acute{\alpha}\nu\omega$ $\gamma\grave{\alpha}\rho\acute{\epsilon}\kappa$ $\pi\alpha\iota\delta\grave{o}\varsigma$ $\acute{\epsilon}\pi\iota\theta\upsilon\mu\tilde{\omega}\nu$ $\kappa\tau\acute{\eta}\mu\alpha\tau\acute{o}\varsigma$ $\tau\upsilon$, $\acute{\omega}\sigma\pi\epsilon\rho$ $\acute{\alpha}\lambda\lambda o\varsigma$ $\acute{\alpha}\lambda\lambda o\upsilon$. $\acute{o}$ $\mu\grave{\epsilon}\nu$ $\gamma\acute{\alpha}\rho$ $\tau\iota\varsigma$ $\acute{\iota}\pi\pi\upsilon\upsilon\varsigma$. $\acute{\epsilon}\pi\iota\theta\upsilon\mu\epsilon\tilde{\iota}$ $\kappa\tau\tilde{\alpha}\sigma\theta\alpha\iota$, $\acute{o}$ $\delta\grave{\epsilon}$ $\kappa\acute{\upsilon}\nu\alpha\varsigma$, $\acute{o}$ $\delta\grave{\epsilon}$ $\chi\rho\upsilon\sigma\acute{\iota}o\nu$, $\acute{o}$ $\delta\grave{\epsilon}$ $\tau\iota\mu\grave{\alpha}\varsigma$: $\acute{\epsilon}\gamma\grave{\omega}$ $\delta\grave{\epsilon}$ $\pi\rho\grave{o}\varsigma$ $\mu\grave{\epsilon}\nu$ $\tau\alpha\tilde{\upsilon}\tau\alpha$ $\pi\rho\acute{\alpha}\omega\varsigma$ $\acute{\epsilon}\chi\omega$, $\pi\rho\grave{\epsilon}o\varsigma$ $\delta\grave{\epsilon}$ $\tau\grave{\eta}\nu$ $\tau\tilde{\omega}\nu$ $\varphi\acute{\iota}\lambda\omega\nu$ $\kappa\tau\tilde{\eta}\sigma\iota\nu$ $\pi\acute{\alpha}\nu\upsilon$ $\acute{\epsilon}\rho\omega\tau\iota\kappa\tilde{\omega}\varsigma$, $\kappa\alpha\grave{\iota}$ $\beta\upsilon\upsilon\lambda o\acute{\iota}\mu\eta\nu$ $\acute{\alpha}\nu$ $\mu\upsilon\iota$ $\varphi\iota\lambda o\nu\acute{\alpha}\gamma\alpha\theta\grave{o}\nu$ $\gamma\epsilon\nu\acute{\epsilon}\sigma\theta\alpha\iota$ $\mu\tilde{\alpha}\lambda\lambda o\nu$ $\grave{\eta}$

τὸν ἄριστον ἐν ἀνθρώποις ὄρτυγα ἢ ἀλεκτρυόνα, καὶ ναὶ μὰ Δία ἔγωγε μᾶλλον ἢ ἵππον τε καὶ κύνα — οἶμαι δέ, νὴ τὸν κύνα, μᾶλλον ἢ τὸ Δαρείου χρυσίον κτήσασθαι δεξαίμην πολὺ πρότερον ἕταιρον, μᾶλλον <δὲ> ἢ αὐτὸν Δαρεῖον — οὕτως ἐγὼ φιλέταιρός τίς εἰμι. ὑμᾶς

All people have their fancies; some desire horses, and others dogs; and some are fond of gold, and others of honour. Now, I have no violent desire of any of these things; but I have a passion for friends; and I would rather have a good friend than the best cock or quail in the world: I would even go further, and say the best horse or dog. Yea, by the dog of Egypt, I should greatly prefer a real friend to all the gold of Darius, or even to Darius himself: I am such a lover of friends as that.

Socrates, in Lysis (Plato)

DEDICATED TO MY FRIEND ANTONIO GRECO WHO INTRO-
DUCED ME TO THE INTRICACIES AND THE BEAUTIES OF
ASYMPTOTIC METHODS IN NONLINEAR WAVE PHENOMENA
(A. M. Anile)

References

1. P. A. Markowich, C. A. Ringhofer and C. Schmeiser, *Semiconductor Equations* (Springer, 1990).
2. C. Jacoboni, P. Lugli, *The Monte Carlo Method for Semiconductor Device Simulations* (Springer, 1989).
3. K. Tomizawa, *Numerical simulation of submicron semiconductor devices* (Artech House, Boston, 1993).
4. J. W. Jerome, *Analysis of charge transport. A mathematical study of semiconductor devices* (Springer, 1995).
5. C. Cercignani, *Rarefied Gas Dynamics* (Cambridge University Press, 2000).
6. J. J. Duderstadt and W. R. Martin, *Transport Theory* (Wiley, New York, 1979).
7. W. M. Stacey Jr., *Variational Methods in Nuclear Reactor Physics* (Academic Press, New York, 1974).
8. A. Schenk, *Advanced Physical Models for Silicon Device Simulation* (Springer, Wien, 1998).
9. G. Alì, A. M. Anile and G. Mascali, *Rendiconti del Circolo Matematico di Palermo*, Serie II, Suppl. **78**, 11-18 (2006).
10. J. J. Sakurai, *Modern Quantum Mechanics* (Benjamin/Cummings, 1985).
11. G. Alì, G. Mascali, *Variational formulation of the steady Boltzmann equation for semiconductors*, in preparation.
12. C. Jacoboni and L. Reggiani, *Rev. Mod. Phys.*, **55**, 645-705 (1983).
13. A. M. Anile, G. Mascali and V. Romano, *Recent developments in hydro-dynamical modeling of semiconductors*, in Lecture Notes in Mathematics, *Mathematical problems in semiconductors*, vol. 1823 (Springer, 2003).

GENERATING MULTI STATE CELLULAR AUTOMATA BY USING CHUA'S "UNIVERSAL NEURON"

Eleonora Bilotta and Gianpiero Di Blasi

Department of Linguistics, University of Calabria,
Arcavacata Rende, 87036, Italy
E-mail: bilotta, gdiblasi@unical.it
www.unical.it

Sebastiano Giambò

Department of Mathematics, University of Messina
Messina, 98166, Italy
E-mail: giambo@mat520.unime.it

Pietro Pantano

Department of Mathematics, University of Calabria,
Arcavacata Rende, 87036, Italy
E-mail: piepa@unical.it
www.unical.it

In 1999 Chua demonstrated that it is possible to obtain boolean Cellular Automata (CA) by using Cellular Neural Networks (CNNs), implemented as chips, which definitely reduce the time of simulation for discrete dynamical systems such as CA and allow for a better understanding of complexity. In this work we demonstrate that Chua's Universal Neuron which simulates boolean CA can be generalized for multistate CA. This new approach allows to investigate the nature of the CA rule space, in relationship with the Universal Neuron parameter space, establishing relationships between discrete and continuous dynamical systems and analysing the nature of local and global rules in affecting the behaviour of these systems. The method which we have developed is fruitful since we have found a lot of CA which can be considered complex rules of class IV, in the Wolfram classification. Furthermore we have used the idea of genetical computation in considering the values of the Universal Neuron parameter space as a sort of genotype which determines and produces variations in the CA behaviour. On this basis, we have implemented a genetic algorithm which fits very well with the searching for CA complex rules.

The nature of the relationship between continuous and discrete systems is a very important topic in the Science of Complexity which helps us to clarify the temporal dimension of many biological phenomena and processes.

1. Introduction

Introduced by von Neumann in the late 1950's[1] Cellular Automata (CA) are one of the paradigms of the Science of Complexity, in the program aimed at investigating the "logic of life" and recreating the structure of life in a synthetic artificial world.[2-4] In fact, the automatic treatment of the self-reproducing process has become a mathematical problem; the "logic of life" or rather, how to extract mathematical algorithmic structures from biological phenomena, structures which can be useful to understand the biology of the phenomena and their possible reinvention in digital worlds. CA are used in many domains of contemporary science from Physics to Biology, from Geology to Social sciences since they represent simple programs which can simulate and explain the emergence of a great quantity of behaviours from simple to complex to chaotic. For a systematic review of this topic see [5].

The nature of CA is discrete since the system is composed of cells that change their state according to the states of their neighbours. Neverthless, the great amount of different behaviours, which CA manifest, can be considered as some typically continuous phenomena, such as wave propagation, which are usually known as solitons.[6] The behaviour of multistate CA (Cellular Automata whose states can assume values different from 0 and 1, especially complex multistate CA) shows surprising analogies with the behaviour of non-linear waves in continuous systems. Furthermore, if we assume that the cell state of a CA is real number, it is possible to use these systems as possible models of partial derivative equations.[5] In the same time, the study of CA behaviour can shed light on wave propagation phenomenon or create a tractable simulation environment of complex processes or consider CA as approximate models of continuous phenomena non tractable with classical methods. In 1999 Chua demonstrated that it is possible to obtain boolean Cellular Automata (CA) by using Cellular Neural Networks (CNNs), implemented as chips which definitely diminishes the time of simulation for discrete dynamical systems as CA and allows for a better understanding of complexity. Introduced by Chua in 1988, CNNs have been used first as pattern recognition systems. They are a collection of cells, connected to other cells in a neighbourhood, where each cell is a small non-linear Chua's circuit. From a technological point of view, CNNs currently represent a method for creating a cell array of meaningful size for image processing and other applications. In fact it is possible to arrange more than 4 million of transistors into a 128 x 128 CNN chip on 1 square centimeter area of silicon.[7] This chip is called the CNN Universal Chip.

14

Their speed in the computational process is 500 times higher than that of other computational systems. CNNs can be used also to study reaction-diffusion processes and non-linear wave propagation phenomena.[8]

Despite their apparent simple criteria of evolution, CA can display rich and complex patterns, whose organization is completely unpredictable. Wolfram[9] classified Cellular automata rules qualitatively according to their asymptotic behaviour: class I (homogeneity); class II (periodicity); class III (chaos); class IV (complexity). Complex rules are important since they perform computation, transmission, storage and modification of information, like biological systems and life. Furthermore, universal computation requires memory and communication over long distance in space and time. Therefore complex computation requires long transient and space-time correlation lengths, typical of complex rules of class IV. Langton found that complex CA rules can be found in this restricted region, called the Edge of Chaos and that the complex dynamics of systems in the vicinity of a phase transition rest on a fundamental capacity for producing information. In some related works[10,11] we have used genetic algorithms for searching complex multistate CA rules. We have found a lot of self-reproducing systems in 2D CA,[12,13] which can be considered as proto-organisms for structure replication that provide insight into analogous processes in the biological world.

In this work, starting from results obtained by Chua[14,15] we demonstrate that multistate CA can be obtained starting from a universal neuron, subject to a process of mutation.

The work is organized as follows. In the second section we introduce formal aspects of CA, while in section 3 we illustrate the method that Chua has used for simulating Boolean CA. In section 4 we generalize this method and apply it to multistate CA. Finally, section 5 contains some examples of generation of complex rules by using this method. Conclusion on the discrete-continuos organization finishes this work.

2. CA formal aspects

The environment considered is a one-dimensional CA, which can be thought as the following tuple:

$$A = (d, \dot{S}, N, f) \tag{1}$$

where d is a positive integer that indicates the CA dimension (one, two, three or more), S a finite set of k states, $N = (x_1, \ldots, x_n)$ is a neighbourhood vector of n different elements of Z^d, f is a local rule defined as:

$$f : S^n \to S \qquad (2)$$

In our case $d = 1$, and the neighborhood identifies the cells with a local interaction ray r, so (2) associates to the $(2r + 1)$ elements of S another element of S, that is

$$(\ldots s_i \ldots) \mapsto s_i \qquad (3)$$

A rule that discriminates all possible cases is expressed in exhaustive form, and considers all $k^{(2r+1)}$ possible cases. In Elementary Cellular Automata (ECA), $S = \{0,1\}$, for each cell i N is composed of the cell itself and of its right and left neighbors. An example of how the local rule (2) works is represented in the following table:

$$\begin{bmatrix} 111 & 110 & 101 & 100 & 011 & 010 & 001 & 000 \\ 1 & 0 & 0 & 0 & 0 & 0 & 0 & 1 \end{bmatrix} \qquad (4)$$

A visual example of table (4) is presented in Figure 1, Table 4 is also called

Fig. 1.   Visual representation of table 4.

CA rule table. A rule table can be respresented in a synthetic form as a sequence of 8 bits in the form riported in (5).

$$l = (10000001) \qquad (5)$$

Each ECA has a local rule which can be represented by using a string of 8 bits and is univocally represented by a number between 0 and 255, which is the decimal corresponding to the string 1.

2.1. *The Universal Neuron*

Following Chua,[14] it is possibile to represent the rule table as a cube whose vertices are white or black (Figure 2). A white or black sign is assigned to each vertex of the boolean cube according to whether the corresponding value of the bit in the string is white or black. A vector u_i with $i = 0, 1, \ldots 7,,$

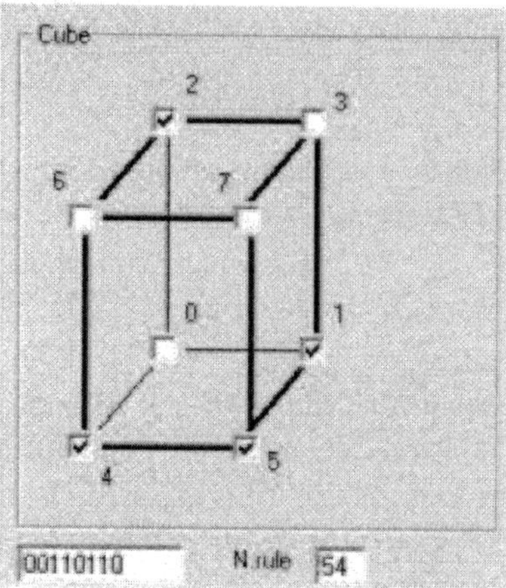

Fig. 2. The string that identifies the ECA rule 54 is visualized in the bottom on the left side of the figure. At each vertex of the boolean cube we have put a black (=0) or white(=1) dot accordingto the string values. Vertex 0 corresponds to the last position in the string; Vertex 7 corresponds to the first position in the string.

corresponds to each vertex of the cube. The components of the vectors are:

$$\begin{cases} u_0 \equiv (-1,-1,-1),\, u_1 \equiv (-1,-1,1),\, u_2 \equiv (-1,1,-1),\, u_3 \equiv (-1,1,1) \\ \;\; u_4 \equiv (1,-1,-1),\;\;\; u_5 \equiv (1,-1,1),\;\;\; u_6 \equiv (1,1,-1),\;\;\; u_7 \equiv (1,1,1) \end{cases} \tag{6}$$

The projection of these vectors on another vector $b \equiv (b_1, b_2, b_3)$, will detect dots marked in white and black, according to the color of the preceding vertex . Figure 3 shows these projections. Following Chua, we indicate by

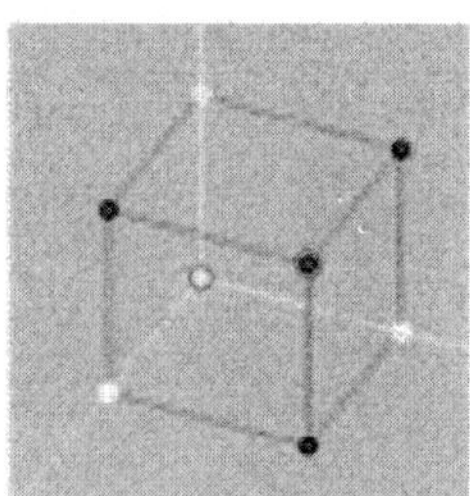 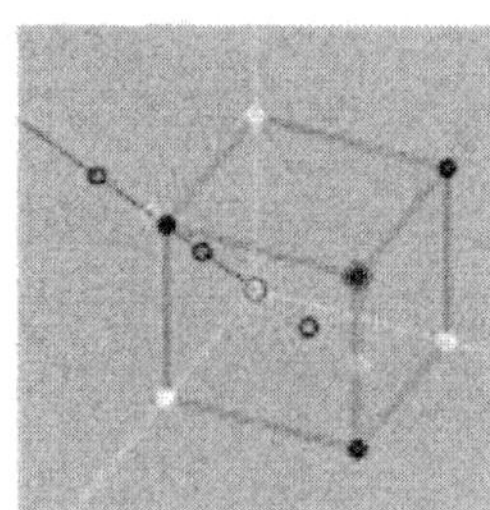 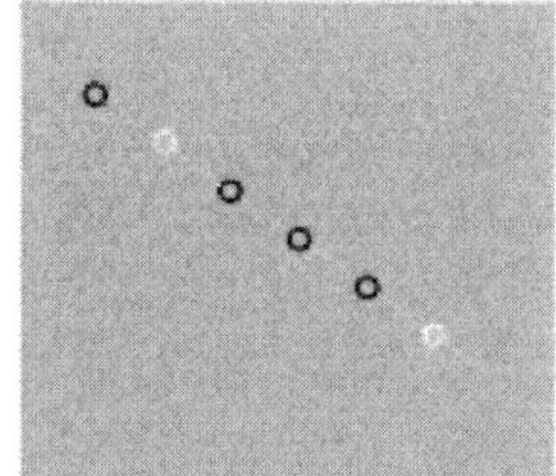

Fig. 3. The vertex of the boolean cube are projected on the vector b.

σ_i the scalar product of b and u_i:

$$\sigma_i = b \cdot u_i \tag{7}$$

It is important to note that it individuates in a precise way the neighborhood configuration in the rule table. Let us build a discriminating function (*off-set level*) in the following way:

$$w(\sigma) = z_2 \pm |[z_1 \pm |z_0 + \sigma|]| \tag{8}$$

where z_0, z_1, z_2 are integers. Consequently, the discriminating function depends on 8 parameters $z_2, z_1, z_0, b_1, b_2, b_3 v$, and two integers which can assume the values ± 1. After having fixed the set of 8 parameters, a value of the discriminating function equal to $w(\sigma_i)$ will consequently correspond to each local rule $i = 0, 1, \ldots 7$.

Let us consider the following dynamical system:

$$\dot{x} = g(x) + w(\sigma_i) \tag{9}$$

where

$$g(x_i) = -x_i + |x_i + 1| - |x_i - 1| \tag{10}$$

is called *driving function*. For each value of σ_i, (9) individuates 8 different dynamical systems. If we consider as initial data of (9) $x(0) = 0$, according to the sign of $w(\sigma)$, (9) will have an unique attractor, positive if $w(\sigma) > 0$, and negative in the opposite case. In Figure 4 some attractors generated by the system (9) for different values of the discriminating function are presented. From what we have discussed above, the following fundamental theorem stems:

- The state $x(t)$ of a dynamical system like (9), with initial condition $x(0) = 0$, converges monotonically toward a positive attractor $Q+$ if $w > 0$ and toward a negative attractor $Q-$ if w< 0.

Let us define the following value y_i as output of the dynamical system (9)

$$y_i = \mathrm{sgn} w(\sigma_i) \tag{11}$$

We have built a machine based on the dynamical system (9) which takes as input a neighbourhood u_i and gives as output y_i. In this way we can generate rule table like table (4) or (5). This machine is ruled by the set of 8 parameters $z_2, z_1, z_0, b_1, b_2, b_3, c_1, c_2$, where c_1 and c_2 are two integers which can assume the values ± 1 (See Figure 5):

$$y_i = \mathrm{sgn} \{z_2 + c_2 |[z_1 + c_1 |z_0 + b \cdot u_i|]|\} \tag{12}$$

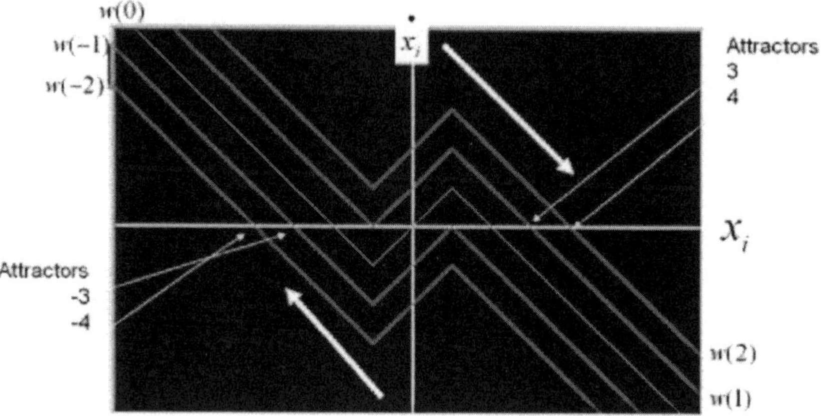

Fig. 4.   In this figure, some attractors generated by the system (9) for different values of the discriminating function are represented.

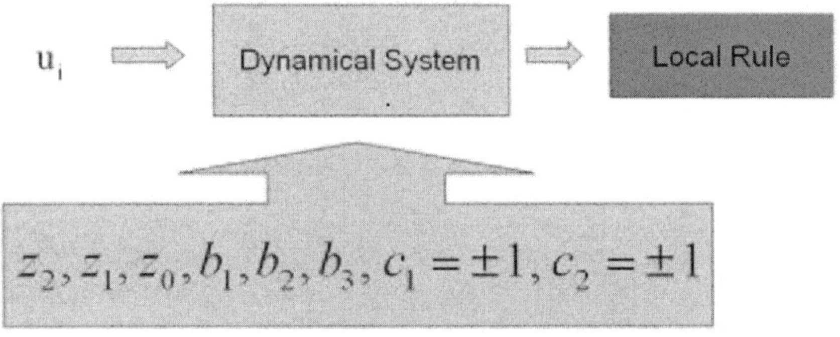

Fig. 5.   The universal neuron computes the data of the neighbourhood and gives as output the corrisponding local rule.

Since dynamical systems like (9) are easily implemented on chips as CNN, it is possibile to create analogous systems that simulate the CA behaviour and definitely reduce the time of simulation for discrete dynamical systems such as CA.[14] Associating parameters to the universal neuron, it is possible to generate the whole set of the ECA rules. In Table 1 some rules and the corresponding universal neuron values are reported. It is important to

Table 1. In this Table, the values of the universal neuron for some ECA are reported .

N. Rule	z_2	c_2	z_1	c_1	z_0	b_1	b_2	b_3
0	1	-1	1	1	2	-1	1	0
1	3	-1	1	1	3	1	1	1
2	2	-1	1	-1	-1	1	1	4
3	2	-1	0	1	3	1	1	0
5	2	-1	3	-1	1	2	0	2
6	4	-1	2	1	3	3	1	1
9	2	-1	0	1	3	3	1	-1
10	3	-1	1	1	2	3	0	1
12	3	-1	2	1	2	3	1	0
13	2	-1	0	1	3	2	-1	1
14	2	-1	0	-1	2	3	1	1
15	2	-1	1	-1	3	3	2	0
16	2	-1	0	-1	3	2	1	4
17	2	-1	4	-1	1	0	3	3
18	1	-1	2	-1	1	2	1	4
20	1	-1	0	-1	2	3	3	2
21	3	-1	0	-1	3	1	1	2
22	1	-1	1	-1	2	1	2	2
24	2	-1	1	1	0	3	1	2
26	3	-1	1	-1	1	4	2	3
32	2	-1	0	1	1	1	4	1
34	3	-1	1	1	2	0	4	1
36	1	-1	0	1	0	-1	1	2

underline that the parameter space of the universal neuron is bigger than the ECA rule space and that to one ECA rule can correspond more than one universal neuron, that is to say more dynamical systems like that presented in (9).

3. Generating multistate CA

The method described above can simply be generalized to the multistate CA. We define multistate cellular automata systems in which the state of the CA cell can assume integer values from 0 to $k-1$, expressed as follows:

$$S = \{0, 1, \ldots, k-1\}. \tag{13}$$

In this case, we cannot use the boolean cube, but we can map the neighbourhood values on specific points of the same cube. The lenght of the string that represents the rule table will be equal to $k^{(2r+1)}$. On the cube it will be possible to detect a corresponding number of points. In Figure 6 two examples of the application of this generalization, with $k = 3$ (on the left) and $k = 4$ (on the right) are showed. The vector u_i with $i = 0, 1, 2, \ldots, (k^{(2r+1)} - 1)$

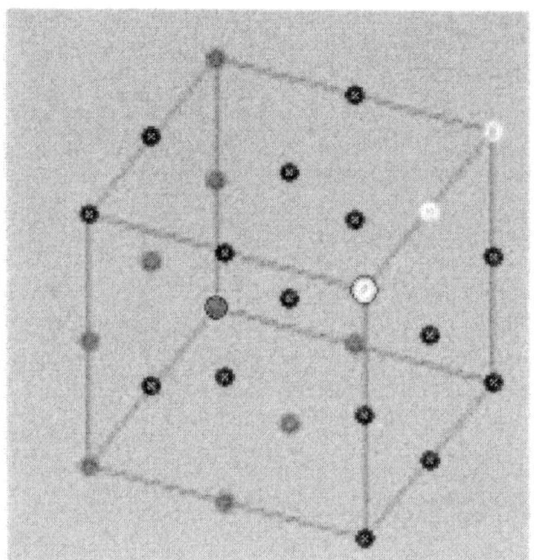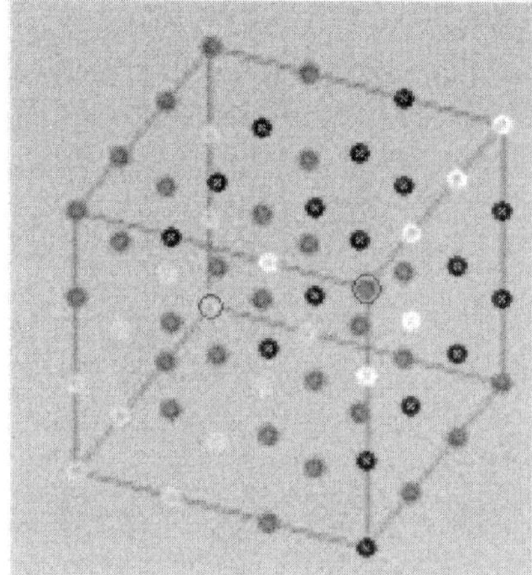

Fig. 6. In this Figure, on the left it is shown a representation of the neighbourhood values for $k = 3$ (the string is: 011112122011112121011112222). On the right a representation of the neighbourhood values for $k = 4$ (the rule is: 20110122122322330012112212232233011211221223223201121122222232333)

has the following components:

$$u = (p, q, r) \tag{14}$$

where $p \in S$, $q \in S$, $r \in S$. So that

$$u_0 = (0,0,0), u_1 = (0,0,1), u_2 = (0,0,2), \ldots, u_{k^{(2r+1)}-1} = (k-1, k-1, k-1)$$

$$w(\sigma) = z_2 + c_2 \left| \left[z_1 + c_1 \left| z_0 + \sigma \right| \right] \right| \tag{15}$$

Also in this case the discriminating function depends on 8 parameters z_2, z_1, z_0, b_1, b_2, b_3, c_2 and c_1, where c_2 and c_1 are two integers which can assume the values ± 1. Furthermore, it is possible to introduce a dynamical system like (9), with the same driving function (10), whose attractors will be fixed points Q_i and they are as many as the neighbours present in the local rule. They will be equal to $k^{(2r+1)}$. In the same way,

- Given a neighbourhood $u(p,q,r)$, the state $x(t)$ of each dynamical system, with initial conditions $x(0) = 0$, converges monotonically toward an attractor Q_{qp}.

In order to calculate the output, we define

$$\bar{y}_{pqr} = (k-1)\frac{Q_{qp} - min}{max - min} \tag{16}$$

where max and min are the maximum and minimus values of the attractors The output y_{pqr} which corresponds to the neighbourhood $u(p,q,r)$ will be then defined as the integer y_{pqr} nearer to $\bar{y}_{pqr}$, In order to calculate the output, we define

$$y_{pqr} = C\mathrm{int}(\bar{y}_{pqr}) \tag{17}$$

In Figure 7, different rules obtained by the same neuron, with different values of k, are visualized, considering the universal neuron that generates the rule 110 with k=2. As it is possible to see, the rules present some emergent structures such as regular domains and gliders. The behaviour of the obtained rules presents many differences as well. This approach is very useful

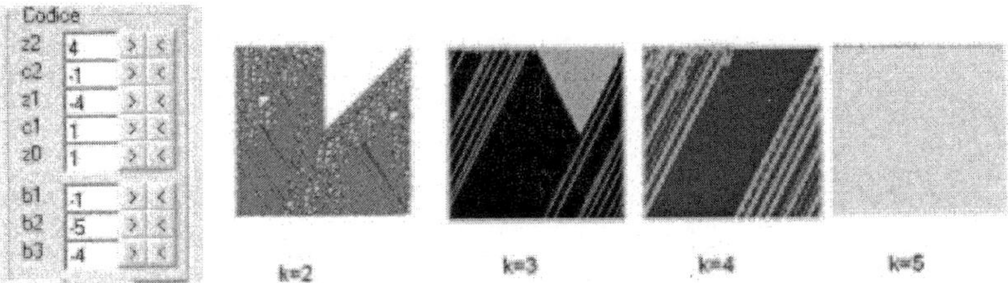

Fig. 7.   In this image, different rules obtained by the same neuron, modifying the value of k are visualized.

in searching for multistate CA rules. Furthermore, the method of searching CA complex rules by using dynamical systems allows the generation of chips endowed with these mechanisms and the use of parallel analogical systems with many advantages with respect to traditional methods.

4. Genetical approach

The method of the universal neuron allows for a higher level of searching for complex rules and the analysis of their corresponding genetic traits. For example, it was possible to investigate the relationship between the change of the universal neuron parameters and the change in the complex rules patterns. Figure 8 shows how CA patterns change, varying parameter b_3 (that is to say changing the orientation of the vector on which the points of the cube are projected). As it is possible to see, the rules present behaviours which go from complex to ordered organizations. In the follow-

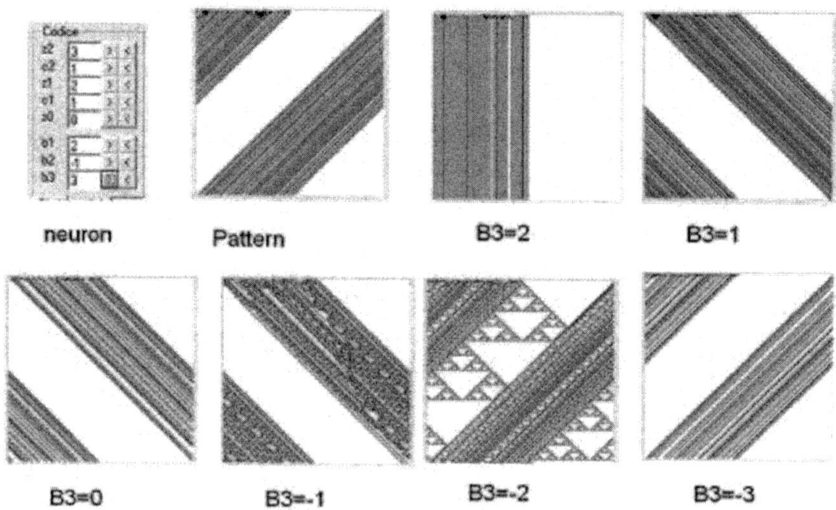

Fig. 8. In this image a collection of patterns that change in relation to the variations of parameter b_3.

ing Table 2, there are the related collection of rules presented in Figure 8 and on the right the values of the universal neuron, with the variation of parameter b_3. From our experiments, the variations of the parameters b1

Table 2. In this table the related collection of rules presented in Figure 8 and on the right the values of the universal neuron, with the variation of parameter b3 are reported.

Rule								
2211321132213221211022102211321121002110211022101101110021002110 >	3	1	2	1	0	2	-1	3
2211222132213222211022102211222111002100211022101001100011002100 >	3	1	2	1	0	2	-1	2
2211222132223322111021112211222110001100111021110011000110001100 >	3	1	2	1	0	2	-1	1
2222222222223333000011112222222200000000000011112222111100000000 >	3	1	2	1	0	2	-1	0
0012012212222231000000100120122210210010000001322222122102100 >	3	1	2	1	0	2	-1	-1
1001101100120112210011011001101122102110210011013221321122102110 >	3	1	2	1	0	2	-1	-2
2101100110011012210021002101100122102210210021003221321022102210 >	3	1	2	1	0	2	-1	-3

and b2, even with negative values, are very interesting for the presence of complex structures in the patterns. The variations of the parameter z is instead less relevant. This new approach allows to investigate the nature of the CA rule space, in relationship with the Universal Neuron parameter space, establishing relationships between discrete and continuous dynamical systems and analysing the nature of local and global rules in affecting the behaviour of these systems.

References

1. von Neumann, J. (1966). *Theory of Self-Reproducing Automata.* University of Illinois Press, Illinois. Edited and completed by A. W. Burks.
2. Langton, C.G. (1984). Self-Reproduction in Cellular Automata. *Physica D*, 10, pp. 135-144.
3. Langton, C.G. (1986). Studying artificial life with cellular automata. *Physica D*, 22, pp. 120-149.
4. Langton, C.G. (1988). Computation at the edge of chaos, *Physica D*, 42, pp. 12-37.
5. Wolfram, S. (2002). *A New Kind of Science.* LLC, Canada.
6. Aizawa Y., Nishikawa I., Kaneko K. (1990) Soliton Turbulence in one-dimensional Cellular Automata, *Physica D*, 45, pp. 307-327
7. Liñan G., Rodriguez-Vazquez A., Espejo S. and Dominguez-Castro R. (2002) "ACE16K: A 128×128 focal plane analog processor with digital I/O," *Proc. 7th IEEE Int. Workshop on Cellular Neural Networks and their Applications*, ed. Tetzlaff, R. (World Scientific, Singapore), pp. 132-139.
8. Chua L.O. (1998). *CNN: A paradigm for complexity*, World Scientific.
9. Wolfram S. (1984). Universality and Complexity in Cellular Automata, *Physica D*, 10, pp. 1-35.
10. Bilotta E., Lafusa A. and Pantano P. (2003). Searching for complex CA rules with GA's. *Complexity*, 8 (3), pp. 56-67
11. Bilotta E., Lafusa A. and Pantano P. (2003). Life-like self-reproducers. *Complexity*, 9 (1), pp. 38-55.
12. Bilotta E. and Pantano P. (2005). Emergent patterning phenomena in 2D Cellular automata. *Artif Life*, 11 (3), pp. 339-362.
13. Bilotta E. and Pantano P. (2006). Structural and Functional Growth in Self-Reproducing Cellular Automata. *Complexity*, 11 (6), pp. 12-29.
14. Chua L.O., Yoon S. and Dogaru R. (2002). A nonlinear dynamics perspective of Wolfram's new kind of science. Part I: Threshold of complexity. *Int. J. Bifurcation and Chaos*, 12, pp. 2655-2766.
15. Chua L.O., Sbitnev V.I. and Yoon S. (2004). "A nonlinear dynamics perspective of Wolfram's new kind of science. Part III: Predicting the unpredictable," *Int. J. Bifurcation and Chaos*, 14, pp. 3689-3820.

ISOCLINE CURVES AND VARIATIONAL SCALAR FIELD

GUY BOILLAT and AUGUSTO MURACCHINI

CIRAM - Research Center of Applied Mathematics, University of Bologna,
Bologna, 40123, Italy.
E-mail: boillat@ciram.unibo.it
E-mail: muracchi@ciram.unibo.it

The equations of isocline curves can be obtained from a variational principle with the *exceptional* scalar field Lagrangian. Shocks are shown to propagate on characteristic curves. Monge-Ampère equation, von Kármán fluid, and Born-Infeld theory appear as examples.

Keywords: Born-Infeld, Monge-Ampère, von Kármán fluid, Characteristic shocks, Euler's equations, Exceptional waves, Hyperbolic systems, Isocline curves.

Al caro amico Antonio
con tanto affetto

1. Isocline curves

It is well known that a completely hyperbolic *exceptional* system[1] of two equations in two variables can be reduced to the simple form of two isocline curves.[2] More generally suppose a field $\mathbf{u}(t, x)$ satisfies a set of N hyperbolic partial differential equations

$$\mathbf{u}_t + A(\mathbf{u})\mathbf{u}_x = 0 \tag{1}$$

together with k constraints[3]

$$C\mathbf{u}_x = 0 \tag{2}$$

compatible only with two waves of velocities λ and λ'.

To the characteristic value λ are associated m linearly independent eigenvectors $\mathbf{d}_a$

$$(A - \lambda I)\,\mathbf{d}_a = 0 \;\; ; \;\; C\,\mathbf{d}_a \equiv 0 \quad (a = 1, 2, ..., m) \tag{3}$$

which are obtained from (1) by the substitution

$$\partial_t \to -\lambda\delta \ , \ \ \partial_x \to \delta \tag{4}$$

so that

$$(A - \lambda I)\delta u = 0 \, .$$

Let $\mathcal{D}$ be the subspace generated by these eigenvectors

$$\mathbf{d} \in \mathcal{D} \ \Rightarrow \ \mathbf{d} = \sum_{a=1}^{m} \alpha^a \mathbf{d}_a \, .$$

Suppose now that λ is an *exceptional*[4] (linearly degenerated) velocity meaning its gradient with respect to $\mathbf{u}$ is orthogonal to the corresponding eigenvectors. This important feature due to Peter Lax[1] is characterized by

$$\delta u = \boldsymbol{\nabla}\lambda\,\mathbf{d} \equiv 0 \tag{5}$$

for any $\mathbf{d} \in \mathcal{D}$ and is already encountered in a number of fields.[3,5] Furthermore assume that *the same properties (3), (5) hold for* λ' (of multiplicity m').

Let us complete the N-dimensional $\mathbf{u}$-space with the remaining eigenvectors of the $N - (m + m') = k$-dimensional space $\overline{\mathcal{D}}$ so that for instance

$$\mathbf{u}_x = \mathbf{d} + \mathbf{d}' + \overline{\mathbf{d}} \ , \ \ \overline{\mathbf{d}} \in \overline{\mathcal{D}}$$

which reduces to

$$\mathbf{u}_x = \mathbf{d} + \mathbf{d}'$$

by effect of (3) and the constraints (2)

$$C\overline{\mathbf{d}} = 0 \ \Rightarrow \ \overline{\mathbf{d}} = 0 \ , \ \ \overline{\mathbf{d}} \in \overline{\mathcal{D}} \, .$$

It follows

$$\partial_t \lambda + \lambda' \partial_x \lambda = 0 \tag{6}$$

since the first member is equal to

$$\boldsymbol{\nabla}\lambda(-A\mathbf{u}_x + \lambda'\mathbf{u}_x) = -\boldsymbol{\nabla}\lambda(A - \lambda'I)\mathbf{u}_x = (\lambda' - \lambda)\boldsymbol{\nabla}\lambda\mathbf{d} = 0 \, .$$

This means that the slope of the λ-curve is constant along the λ'-curve and vice versa[2]

$$d\lambda/dt = 0 \ , \ \ dx/dt = \lambda' \, .$$

26

2. Von Kármán-Tsien fluid

As a first example consider the von Kármán-Tsien fluid[6–8]

$$\partial_t \rho + \partial_x v = 0$$

$$\partial_t v + \partial_x \left(\frac{v^2}{\rho} - p \right) = 0$$

where $p = p_o - (a^2/\rho)$. Wave velocities easily follow by the substitution (4) which yields

$$\delta v - \lambda \delta \rho = 0$$

$$\left(2\frac{v}{\rho} - \lambda \right) \delta v + \frac{1}{\rho^2}(a^2 - v^2)\delta \rho = 0$$

and

$$\lambda^{\pm} = \frac{v \pm a}{\rho}.$$

Exceptionality is easily checked

$$\delta\lambda^{\pm} = \frac{1}{\rho}(\delta v - \lambda^{\pm}\delta\rho) = 0.$$

Therefore

$$\partial_t \lambda^{\pm} + \lambda^{\mp}\partial_x\lambda^{\pm} = 0.$$

3. Monge-Ampère equation

The nonlinear system of N equations

$$u_t + f(u_x) = 0 \tag{7}$$

differentiated with respect to t and x, gives the quasi-linear system of $2N$ equations in the form (1)

$$v_t + F(w)v_x = 0, \quad w_t - v_x = 0 \tag{8}$$

where $v = u_t$, $w = u_x$ and the square matrix F is the derivative of f with respect to w together with the N constraints

$$v_x + F(w)w_x = 0.$$

Assume that F is regular and has only two *exceptional* eigenvalues μ, μ' of multiplicities m, $m' = N - m$. The wave velocities follow from the equations (8)

$$(F - \lambda I)\delta v = 0, \quad \delta v + \lambda \delta w = 0.$$

It appears immediately that $\lambda = 0$ is a solution (due to the reduction (8) to the first order of the differentiated system) which is incompatible with the constraints

$$F\delta w + \delta v = 0 \,.$$

Therefore $\mu(w)$ and $\mu'(w)$ satisfy the isocline equation (6).

As an example consider the Monge-Ampère equation[3] with constant coefficients

$$H\psi_{tt} + 2K\psi_{tx} + L\psi_{xx} + M + N(\psi_{tt}\psi_{xx} - \psi_{tx}^2) = 0 \,. \tag{9}$$

Putting $\psi_t = p$, $\psi_x = q$ we can write

$$p_t + \frac{2Kp_x + Lq_x + M - Np_x^2}{H + Nq_x} = 0$$

$$q_t - p_x = 0$$

which is just of the form (7) with $\mathbf{u} = (p, q)$. The eigenvalues $\mu(w)$ and $\mu'(w)$ are easily seen to be the roots of the characteristic equation

$$\lambda^2(H + Nq_x) - 2\lambda(K - Np_x) + L + Np_t = 0 \,.$$

Taking account of (9) they assume the simple expression

$$\frac{K - Np_x \pm \sqrt{K^2 - LH + MN}}{H + Nq_x}$$

and satisfy (6).

4. Born-Infeld nonlinear electrodynamics

The equations of nonlinear electrodynamics[9]

$$\partial_t \vec{D} - \operatorname{curl} \vec{H} = 0$$

$$\partial_t \vec{B} + \operatorname{curl} \vec{E} = 0$$

depend on the nonlinear function $L(Q, R)$ of the electromagnetic invariants

$$Q = (\vec{B}^2 - \vec{E}^2)/2 \,, \quad R = \vec{E} \cdot \vec{B}$$

through

$$\vec{D} = L_Q \vec{E} - L_R \vec{B} \,, \quad \vec{H} = L_Q \vec{B} + L_R \vec{E} \,.$$

They are completed by the constraints

$$\operatorname{div} \vec{D} = 0 \,, \quad \operatorname{div} \vec{B} = 0$$

which eliminate a double stationary wave ($\lambda = 0$). In general four distinct velocities subsist.

Only in the Born-Infeld case[10,11]

$$L = \sqrt{-R^2 + k(2Q + k)} \ , \quad k = \text{constant} > 0$$

two double (and therefore *exceptional*) velocities are found. In one dimensional propagation they are the roots of the polynomial[12]

$$(\vec{B}^2 + k)\lambda^2 - 2\lambda S_1 + \vec{E}^2 - k - E_1^2 - B_1^2 = 0 \ , \qquad \vec{S} = \vec{E} \times \vec{B}$$

each of which satisfies an equation of type (6).

5. Conservation laws

The system of isocline curves for which explicit solutions[13] and weak solutions have been given[14–17]

$$\partial_t u + v\partial_x u = 0$$

$$\partial_t v + u\partial_x v = 0$$

admits an infinity of conservation laws of the form

$$\partial_t \frac{f(u) - g(v)}{u - v} + \partial_x \frac{vf(u) - ug(v)}{u - v} = 0 \ . \tag{10}$$

However all these expressions are not equivalent when weak solutions are concerned. In fact, in this case, the Rankine-Hugoniot conditions must replace each equation of the original differential system

$$\left[\frac{vf(u) - u\,g(v)}{u - v} - s\,\frac{f(u) - g(v)}{u - v} \right] = 0$$

$$\left[\frac{vF(u) - u\,G(v)}{u - v} - s\,\frac{F(u) - G(v)}{u - v} \right] = 0$$

where s is the shock velocity and the bracket denotes the jump of the field across the shock i.e. $[X] = X_1 - X_0$. Depending on the choice of the various functions f, g, F, G the shocks may or may not occur on characteristic surfaces. This means that besides characteristic shocks for which $s = u$ or $s = v$ other shocks may exist which propagate with a different velocity. In the latter case the system is *completely exceptional* and *strictly exceptional* (only characteristic shocks) in the former case.[18]

Now since all these formulations are not equivalent for weak (discontinuous) solutions a question arises: which conservation laws must be chosen ? While this kind of problem is not likely to appear in mathematical physics where the equations are already balance laws nevertheless we propose to look for a variational principle.

6. Euler's equations

We consider the system of Euler's equations

$$\partial_t P + \partial_x Q = 0 \tag{11}$$

$$\partial_t q - \partial_x p = 0 \tag{12}$$

$$L = L(\Psi_t, \Psi_x) = L(p, q), \quad p \doteq \Psi_t, \ q \doteq \Psi_x, \ P \doteq L_p, \ Q \doteq L_q \, .$$

Then, from (10),

$$p = \frac{ug(v) - vf(u)}{u - v} \ , \quad q = \frac{f(u) - g(v)}{u - v} \tag{13}$$

$$P = \frac{F(u) - G(v)}{u - v} \ , \quad Q = \frac{vF(u) - uG(v)}{u - v} \, . \tag{14}$$

By differentiation

$$\frac{\partial p}{\partial u} = v\frac{q - f'}{u - v} \ , \quad \frac{\partial p}{\partial v} = -u\frac{q - g'}{u - v} \ , \quad \frac{\partial q}{\partial u} = \frac{f' - q}{u - v} \ , \quad \frac{\partial q}{\partial v} = \frac{q - g'}{u - v}$$

we obtain

$$dL = Pdp + Qdq = (Q - vP)\frac{\partial q}{\partial u}du + (Q - uP)\frac{\partial q}{\partial v}dv = -G\frac{\partial q}{\partial u}du - F\frac{\partial q}{\partial v}dv$$

and therefore

$$\frac{\partial L}{\partial u} = -G\frac{\partial q}{\partial u}, \quad \frac{\partial L}{\partial v} = -F\frac{\partial q}{\partial v} \, .$$

Integrating results in

$$L + Gq = \beta(v), \quad L + Fq = \alpha(u) \tag{15}$$

from which

$$(F - G)q = \alpha - \beta \rightarrow (F - G)(f - g) = (\alpha - \beta)(u - v) \tag{16}$$

i.e.

$$Ff - \alpha u + Gg - \beta v = Fg + Gf - (\alpha v + \beta u) \, .$$

By differentiating with respect to u and then v we have

$$F'g' + G'f' - \alpha' - \beta' = 0 \tag{17}$$

and

$$F''g'' + G''f'' = 0 \, .$$

By supposing $f'' \neq 0$, and since f, g need to be defined up to the same multiplicative constant we can take

$$F'' = -f'', \quad G'' = g''$$

from which

$$F = -f + a_1 u + a_2 , \quad G = g + b_1 v + b_2 \tag{18}$$

with arbitrary constants of integration. Substituting in (17) we get successively

$$\beta' - a_1 g' = b_1 f' - \alpha' = a$$

$$\alpha = b_1 f - au - c_1, \quad \beta = a_1 g + av + c_2 . \tag{19}$$

Then (16) implies

$$(F - b_2)f + (c_2 - \alpha)u = (a_2 - G)g + (c_1 + \beta)v = const. = -c .$$

Replacing F and G by their expressions results in

$$S(f, u) \equiv -f^2 + [(a_1 - b_1)u + a_2 - b_2]f + au^2 + (c_1 + c_2)u + c = 0 \tag{20}$$

$$S(g, v) \equiv -g^2 + [(a_1 - b_1)v + a_2 - b_2]g + av^2 + (c_1 + c_2)v + c = 0 \tag{21}$$

whose solutions are

$$f = -\frac{1}{2}[(b_1 - a_1)u + b_2 - a_2] - \frac{\epsilon_1}{2}\sqrt{R(u)} , \quad \epsilon_1^2 = 1$$

$$g = -\frac{1}{2}[(b_1 - a_1)v + b_2 - a_2] - \frac{\epsilon_2}{2}\sqrt{R(v)} , \quad \epsilon_2^2 = 1$$

and thus, by (18),

$$F = \frac{1}{2}(a_1 + b_1)u + \frac{1}{2}(a_2 + b_2) + \frac{\epsilon_1}{2}\sqrt{R(u)}$$

$$G = \frac{1}{2}(a_1 + b_1)v + \frac{1}{2}(a_2 + b_2) - \frac{\epsilon_2}{2}\sqrt{R(v)}$$

where R is an arbitrary polynomial of order two. From (13), (14), (18) follow

$$p = \frac{ug - vf}{u - v} = \frac{\epsilon_1 v\sqrt{R(u)} - \epsilon_2 u\sqrt{R(v)}}{2(u - v)} - \frac{b_2 - a_2}{2} \tag{22}$$

$$q = \frac{f - g}{u - v} = \frac{\epsilon_2\sqrt{R(v)} - \epsilon_1\sqrt{R(u)}}{2(u - v)} - \frac{b_1 - a_1}{2} \tag{23}$$

$$P = \frac{-f + a_1 u + a_2 - g - b_1 v - b_2}{u - v} = \frac{\epsilon_1 \sqrt{R(u)} + \epsilon_2 \sqrt{R(v)}}{2(u-v)} + \frac{a_1 + b_1}{2}$$

$$Q = \frac{v(-f + a_1 u + a_2) - u(g + b_1 v + b_2)}{u-v} = \frac{\epsilon_1 v \sqrt{R(u)} + \epsilon_2 u \sqrt{R(v)}}{2(u-v)} - \frac{a_2 + b_2}{2} \; .$$

Without loss of generality, the additional constants may be taken equal to zero which means that $a_1 = a_2 = b_1 = b_2 = 0$ and by (20), (21)

$$f^{\,2} = R(u) = au^2 + 2bu + c \; , \quad g^2 = R\,(v) = av^2 + 2bv + c, \quad 2b = c_1 + c_2 \; .$$

7. The scalar field Lagrangian L(p,q)

From (15), (18), (13) and (19)

$$2L = \alpha + \beta - q(F + G) = (q^2 - a)(u - v) + c_2 - c_1 \; .$$

To find the expression of the Lagrangian as a function of p and q we first obtain from (13)

$$(p + uq)^2 = f^{\,2} \Rightarrow u^2(q^2 - a) + 2u(pq - b) + p^2 - c = 0 \; .$$

It is easy to see that v satisfies the same polynomial equation, so that $u - v$ is the difference of its roots and

$$L = \sqrt{2Q + k} \; , \quad Q \doteq \frac{1}{2}(ap^2 - 2bpq + cq^2), \quad k \doteq b^2 - ac$$

since the Lagrangian needs to be defined up to a multiplicative and additive constant. For L to exist for constant field ($\Psi = \mathrm{const.}$) k must be positive. This implies that the quadratic form Q is not positive definite and L is just the *scalar field Lagrangian*. By a special choice of the coordinates we can take

$$L = \sqrt{1 + q^2 - p^2} \; . \tag{24}$$

From (11), (12) the wave velocity λ is easily calculated by solving the system

$$\lambda \delta q + \delta p = 0 \; , \quad \delta Q - \lambda \delta P = \lambda \delta \left(\frac{p}{L}\right) + \delta \left(\frac{q}{L}\right) = 0$$

and satisfies

$$(\lambda p + q)^2 - L^2(1 - \lambda^2) \equiv (p + \lambda q)^2 - (1 - \lambda^2) = 0 \tag{25}$$

which implies $\lambda^2 \leq 1$. (Instead $\lambda^2 \geq 1$ with the choice $L = \sqrt{1 + p^2 - q^2}$). With (22), (23) it is immediate to check that $\lambda = u$, $\lambda = v$ are indeed the roots of this polynomial.

8. Characteristic shocks

The conservative form (11) and (12) yields the Rankine-Hugoniot conditions[6] for the jumps $[p] = p_1 - p_0$ of p and q across the shock front moving with speed s

$$[p + sq] = 0$$

$$[Q] - s[P] = 0 \Rightarrow \left[\frac{ps + q}{L}\right] = 0.$$

The identity (25) gives

$$\frac{(ps + q)^2}{L^2} - (1 - s^2) = \frac{(p + sq)^2}{L^2} - \frac{(1 - s^2)}{L^2}$$

and taking the jump

$$\{(p + sq)^2 + s^2 - 1\} \left[\frac{1}{L^2}\right] = 0$$

from which we obtain the characteristic shocks whose velocities satisfy (25)

$$s = -\frac{pq \pm L}{1 + q^2}$$

where the continuous value of s can be indifferently computed on either side of the shock front (i.e. with $p = p_0$, $q = q_0$ or $p = p_1$, $q = q_1$).

Now assume

$$(p + sq)^2 + s^2 - 1 \neq 0 \tag{26}$$

L (which is positive) is continuous and

$$[p + sq] = 0, \quad [ps + q] = 0, \quad [p^2 - q^2] = [(p + q)(p - q)] = 0.$$

Adding and subtracting the first equations results in

$$(1 + s)[p + q] = 0, \quad (1 - s)[p - q] = 0$$

which implies $[p] = [q] = 0$ if $s^2 \neq 1$.

If $s = \pm 1$, $[p \pm q] = 0$, $(p \pm q)[p \mp q] = 0 \Rightarrow [p \mp q] = 0$ by (26) and $[p] = [q] = 0$. Only characteristic shocks may exist. The system (11), (12), (24) is *strictly exceptional.*

9. Differential equation for λ

From (6), inserting

$$\lambda' = -\frac{\partial_t \lambda}{\partial_x \lambda}$$

into the other isocline equation yields

$$\partial_t \left(\frac{\partial_t \lambda}{\partial_x \lambda} \right) + \lambda \partial_x \left(\frac{\partial_t \lambda}{\partial_x \lambda} \right) = 0$$

and λ satisfies the second order differential equation

$$\lambda_x \lambda_{tt} + (\lambda \lambda_x - \lambda_t) \lambda_{tx} - \lambda \lambda_t \lambda_{xx} = 0$$

which cannot be obtained from a Lagrangian $L(\lambda_t, \lambda_x, \lambda)$. Neither is it possible to find a Lagrangian $L(\lambda_t, \lambda_x, \lambda'_t, \lambda'_x)$ for

$$\lambda_{tt} - \left(\frac{\lambda'_t}{\lambda'_x} + \frac{\lambda_t}{\lambda_x} \right) \lambda_{tx} + \frac{\lambda'_t}{\lambda'_x} \frac{\lambda_t}{\lambda_x} \lambda_{xx} = 0$$

and the analogous equation obtained by exchanging λ and λ' according to a study of systems of the form[19]

$$\mathbf{u}_{tt} + \rho(\mathbf{u}_t, \mathbf{u}_x)\mathbf{u}_{tx} + \nu(\mathbf{u}_t, \mathbf{u}_x)\mathbf{u}_{xx} = 0$$

connected to relativistic strings.[20,21]

References

1. P.D. Lax, *Ann. Math. Studies* **33**, 211 (1954); *Comm. Pure Appl. Math.* **10**, 537 (1957); *Selected Papers*, 2 vol. (Springer-Verlag, New York, 2005).
2. P.D. Lax, *J. Math. Phys.* **5**, 611 (1964).
3. G. Boillat, *Nonlinear Hyperbolic Fields & Waves* in: *Recent Mathematical Methods in Nonlinear Wave Propagation* ed. T. Ruggeri (C.I.M.E. Course, Montecatini Terme, Italy 1994). Lecture Notes in Mathematics, **1640**, (Springer-Verlag, New York, 1996).
4. A. Jeffrey, *Quasilinear hyperbolic systems and waves* (Pitman, London, 1978).
5. G. Boillat and A. Muracchini, *Atti Accad. Peloritana Pericolanti. Cl. Sci. Fis. Mat. Nat.* **LXXX**, 159 (2002).
6. R. Courant and K.O. Friedrichs, *Supersonic Flow and Shock Waves* (Wiley-Interscience, New York, 1948; Springer Verlag, New York, 1999).
7. A. Greco, On the strict exceptionality for a subsonic flow, in II^0 *Congresso Nazionale AIMETA*, **4**, 127 (1974) (Napoli, Italy).
8. N. Manganaro and G. Valenti, *Atti Sem. Mat. Fis. Univ. Modena* **XXXVIII**, 109 (1990).
9. G.W. Gibbons and C.A.R. Herdeiro, *Phys. Rev. D* **63**, 064006-1 (2001); hep-th/0008052.

10. M. Born and L. Infeld, *Nature* **132**, 970 (1933); *Proc. Roy. Soc. A* **144**, 425 (1934).
11. M. Born, *Ann. Inst. Henri Poincaré* **7**, 155 (1937).
12. G. Boillat, *Ann. Inst. Henri Poincaré* **5**A, 217 (1966); *J. Math. Phys.* **11**, 941 (1970).
13. T. Ruggeri, *Riv. Mat. Univ. Parma* (4) **5**, 415 (1979).
14. D. Serre and I. N. Sneddon, *Systems of Conservation Laws,* 2 vol. (Cambridge University Press, 1999-2000).
15. G.Q. Chen, *Arch. Rat. Mech. Anal.* **121**, 131 (1992).
16. Y.- J. Peng, *Appl. Math. Lett.* **11**, 75 (1998).
17. T.-T. Li and Y.-J. Peng, *Nonlinear Anal.* **52**, 573 (2003).
18. G. Boillat and T. Ruggeri, *Bollettino U.M.I.* (5), **15**-A, 197 (1978).
19. T. Ruggeri, *Lett. Nuovo Cimento* **22**, 69 (1978).
20. T. Ruggeri and A. Strumia, *Prog. Theor. Phys.* **59**, 2121 (1978).
21. G. Boillat and T. Ruggeri, *Prog. Theor. Phys.* **60**, 1928 (1978).

Acknowledgments

This paper was supported by MIUR PRIN (Progetto di Ricerca di Interesse Nazionale: *Nonlinear Propagation and Stability in Thermodynamical Processes of Continuous Media.* Coordinator: Prof. Tommaso Ruggeri) and by GNFM-INdAM.

FOKKER-PLANCK ASYMPTOTICS
AND
THE RICCI FLOW

M. CARFORA*

Dipartimento di Fisica Nucleare e Teorica, Universita' degli Studi di Pavia,
and
Istituto Nazionale di Fisica Nucleare, Sezione di Pavia,
Pavia, Via Bassi 6, 27100, Italy
** E-mail: mauro.carfora@pv.infn.it*

We discuss the interplay between Fokker-Planck dynamics and some elementary aspects of the theory of Hamilton-Perelman Ricci flow. In particular, we show that the Fokker-Planck equation is the natural diffusion process for absolutely continuous probability measures (backward) evolving along the Ricci flow. A few preliminary results on the asymptotic trend to equilibrium for such a process are reported.

Keywords: Ricci flow, Relative entropy, Fokker-Planck diffusion.

1. INTRODUCTION

It is a pleasure to dedicate this paper to Antonio, explorer of many lands.

Since the seminal paper by R. Hamilton on Ricci flow[1] appeared in 1982, there has been a great deal of activity in such a field, giving new insight into the geometrical structure of three-manifolds. Let us recall that the Ricci flow is a weakly-parabolic diffusion-reaction equation which deforms the metric g of a Riemannian manifold (Σ, g) in the direction of its Ricci curvature $Ric(g)$. Locally, (*e.g.*, in harmonic coordinates), it can be considered as a heat equation for the metric tensor, or, more formally, as the flow associated with the natural vector field defined by the Ricci tensor on the infinite-dimensional space of Riemannian structure.[2] Under suitable conditions it provides the natural technique for addressing analytically Thurston's geometrization conjecture, uniformizing Riemannian three-manifolds (Σ, g) to specific locally homogeneous model geometries, (see[3-5] for reviews). Not surprisingly, the realization of such a program has proven to be a highly non-trivial task and the recent results by G. Perelman[6] have provided a breakthrough taking the whole subject by storm and opening up to results and techniques of vast potential use also in theoretical and mathematical physics (*e.g.* in relativistic cosmology,[7,8] and quantum field theory[9–11]). In this talk, we shall focalize on an important step in Perelman's strategy: the coupling of the Ricci flow $\beta \mapsto (\Sigma, g(\beta))$ with the (backward) diffusion of a reference probability measure $d\varpi(\beta)$. Such a coupling is so devised as to connect the localization properties of $d\varpi(\beta)$ with the evolution of curvature along the Ricci flow. In this way one can probe the development of singularities along the Ricci flow trajectories by means of a sophisticated energy-entropy argument. A fact, this latter, that has been put to use by Perelman[6,12,13] with great ingenuity in his work on the proof of Thurston geometrization conjecture. It has been often stressed[6,14–16] that Perelman's coupling is strictly related to the theory of logarithmic Sobolev inequalities,[6,17] a typical topic in controlling the trend to equilibrium of diffusive processes. However, it does not seem to have been equally realized that Perelman's coupling directly calls into play Fokker-Planck diffusions as the process relating the fluctuations of the scalar curvature, along the Ricci flow, to the localization behavior of $d\varpi(\beta)$. Actually, it is not difficult to prove[18] that a Fokker-Planck process is the natural diffusion of probability measures $d\Omega_\beta$ along the (volume-normalized) Ricci flow, and that it is associated with a gradient flow generated by the (weakly) monotonic relative entropy $S[d\Omega_\beta \| d\Pi_\beta] \doteq \int_{\Sigma_\beta} d\Omega_\beta \ln \frac{d\Omega_\beta}{d\Pi_\beta}$, (where $d\Pi_\beta \doteq Vol^{-1}[\Sigma_\beta]d\mu_{g(\beta)}$ denotes the normalized Riemannian volume element of $(\Sigma, g(\beta))$). The relevance of these remarks on the role of Fokker-Planck diffusion in Ricci flow theory does not lie much in providing yet another way of discussing Perelman's analysis but rather in the fact that it directly calls into play in such a setting the role of the Wasserstein metric.[19] The distance induced by this metric provides a way of turning the space of probability measures on a Riemannian manifold into a geodesic space, and has recently drawn attention in attempts of extending the notion of Ricci curvature to general metric spaces.[19–22] Thus, one has here a strong indication that the use of the geometry of the space of probability measures and of the associated notions of optimal transport and Wasserstein metric may also play a significant role in extending Ricci flow theory to metric spaces more general than Riemannian manifolds.

The present paper is based on[18] and on my talk at the meeting *Asymptotic Methods in Non-linear Wave Phenomena*, celebrating A. Greco. Reasons of space do not allow me to insert here the (too) many illustrations that accompanied the talk and that may help in visualizing the subtle geometry of the Hamilton-Perelman flow. The web-oriented reader can find the full presentation posted at http://www.pv.infn.it/∼ carfora/ .

2. Probability diffusion and Ricci flow

To set notation, let $d\mu_g$, $Vol\,[\Sigma]_g$, and $d\Pi \doteq Vol\,[\Sigma]_g^{-1}\,d\mu_g$ respectively denote the Riemannian density, the volume, and the corresponding normalized measure on a smooth three-dimensional compact Riemannian manifold (Σ, g) without boundary. Let $Prob(\Sigma)$ be the space of all Borel probability measure on Σ, endowed with the topology of weak convergence, and defined by

$$Prob(\Sigma) \doteq \left\{ N \in C_b(\Sigma, \mathbb{R}^+), \int_\Sigma N\, d\Pi = 1 \right\}, \tag{1}$$

where $C_b(\Sigma, \mathbb{R}^+)$ is the space of positive bounded measurable functions.[23] As suggested by F. Otto,[24] (see also[25]), when discussing probability diffusion semigroups on manifolds it can be profitable to consider $Prob(\Sigma)$ as an infinite dimensional manifold locally modelled over the Hilbert space completion of the tangent space

$$T_N Prob(\Sigma) \doteq \left\{ h \in C_b(\Sigma, \mathbb{R}), \int_\Sigma h\, N d\Pi = 0 \right\}, \tag{2}$$

with respect to the inner product defined, for any φ, $\zeta \in C_0^\infty(\Sigma, \mathbb{R})$, by the Dirichlet form

$$\langle \varphi, \zeta \rangle_N \doteq \int_\Sigma g^{ik}\, \nabla_k \varphi\, \nabla_i \zeta\, N\, d\Pi. \tag{3}$$

Under such an identification, one can represent vectors in $T_N Prob(\Sigma)$ as the solutions of an elliptic problem naturally associated with the given probability measure according to

$$(h, N) \in T_N Prob(\Sigma) \times Prob(\Sigma) \longmapsto \psi \in C_b(\Sigma, \mathbb{R})/\sim, \tag{4}$$

where, for any given pair (h, N), the function ψ is formally determined by the elliptic PDE

$$-\nabla^i\left(N\, \nabla_i \psi\right) = h, \tag{5}$$

under the equivalence relation $\sim$ identifying any two such solutions differing by an additive constant. In general, such a characterization is somewhat heuristic, at least in the sense that its validity must be checked case by case.

Recall that given two probability measures $d\varpi$ and $d\theta \in Prob(\Sigma)$, the relative entropy functional (of $d\varpi$ with respect to $d\theta$) is defined[23] according to

$$
S\left[d\varpi \parallel d\theta\right] \doteq
\begin{cases}
\int_\Sigma \frac{d\varpi}{d\theta} \ln \frac{d\varpi}{d\theta}\, d\theta & \text{if} \quad d\varpi << d\theta \\[2mm]
\infty & \text{otherwise,}
\end{cases}
\tag{6}
$$

where $d\varpi << d\theta$ stands for absolute continuity. Jensen's inequality implies that $S[d\varpi \parallel d\theta] \in [0, +\infty]$. Roughly speaking $S[d\varpi \parallel d\theta]$ provides the rate functional for the large deviation principle[23] controlling how deviant is the distribution of $d\varpi$ with respect to the reference measure $d\theta$ in $Prob(\Sigma)$. In particular one has[23] (Pinsker's inequality)

$$
S[d\varpi \parallel d\theta] \geq \frac{1}{2} \|d\varpi - d\theta\|_{var}^2 \, ,
\tag{7}
$$

where we have introduced the total variation norm on $Prob(\Sigma)$ defined by

$$
\|d\varpi - d\theta\|_{var} \doteq
\tag{8}
$$

$$
\doteq \sup_{\|f\|_b \leq 1} \left\{ \left| \int_\Sigma f\, d\varpi - \int_\Sigma f\, d\theta \right| : f \in C_b(\Sigma; \mathbb{R}) \right\},
$$

with $\|f\|_b \leq 1$ the uniform norm on $C_b(\Sigma; \mathbb{R})$. This is a particular (and elementary) case of transportation inequalities involving $S[d\varpi \parallel d\theta]$ and the notion of Wasserstein distance between probability measures.[26,27] Let us recall that we define the Wasserstein distance of order s between $d\varpi$ and $d\theta$ in the space of probability measures (with finite s-moments) $Prob(\Sigma \times \Sigma)$ as

$$
D_s^W(d\varpi, d\theta) \doteq \inf_{\pi \in I(d\varpi, d\theta)} \left(\iint_{\Sigma \times \Sigma} d(x, y)^s \pi(d\varpi d\theta) \right)^{\frac{1}{s}}
\tag{9}
$$

where $I(d\varpi, d\theta) \subset Prob(\Sigma \times \Sigma)$ denotes the set of probability measures on $\Sigma \times \Sigma$ with marginals $d\varpi$ and $d\theta$, i.e., such that $\pi(\Omega \times \Sigma) = d\varpi(\Omega)$ and $\pi(\Sigma \times \Omega) = d\theta(\Omega)$ for any measurable set $\Omega \subset \Sigma$; ($I(d\varpi, d\theta)$ is often called the set of couplings between $d\varpi$ and $d\theta$). Note that by Kantorovich-Rubinstein duality, we have that $D_1^W(d\varpi, d\theta) = \|d\varpi - d\theta\|_{var}$. Intuitively, $D_s^W(d\varpi, d\theta)$ represents, as we consider all possible couplings between the measures $d\varpi$ and $d\theta$, the minimal cost needed to transport $d\varpi$ into $d\theta$ provided that the cost to transport the point x into the point y is given by $d(x, y)^s$. The distance $D_s^W(d\varpi, d\theta)$ metrizes $Prob(\Sigma)$ turning it into a geodesic space.

The pair $(Prob(\Sigma), D_s^W(d\varpi, d\theta))$ has recently drawn attention[19–22,28] as a suitable setting for extending the notion of Ricci curvature to general metric spaces. On a compact Riemannian manifold we always have the bound ([29] case 5 of Th.1)

$$D_s^W(d\varpi, d\theta) \leq 2^{\frac{1}{2s}} diam(\Sigma, g) S[d\varpi \parallel d\theta]^{\frac{1}{2s}}, \tag{10}$$

where $diam(\Sigma, g) \doteq \sup\{d_g(x, y); x, y \in (\Sigma, g)\}$ denotes the diameter of (Σ, g).

2.1. *The Perelman volume-normalized Ricci flow*

Let us consider the volume normalized Ricci flow $\beta \mapsto g_{ab}(\beta)$, $0 \leq \beta < T^{1,4}$ associated with a metric g_{ab} on a three-dimensional manifold Σ

$$\begin{cases} \frac{\partial}{\partial \beta} g_{ab}(\beta) = -2R_{ab}(\beta) + \frac{2}{3} g_{ab}(\beta) \langle R(\beta) \rangle_{\Sigma_\beta}, \\ \\ g_{ab}(\beta = 0) = g_{ab}, \end{cases} \tag{11}$$

where $R_{ab}(\beta)$ denotes the components of the Ricci tensor of $g_{ab}(\beta)$, and

$$\langle R(\beta) \rangle_{\Sigma_\beta} \doteq \frac{\int_{\Sigma_\beta} R(\beta) d\mu_{g(\beta)}}{[Vol(\Sigma, g_{ab}(\beta))]} \tag{12}$$

is the averaged scalar curvature with respect to the Riemannian measure $d\mu_{g(\beta)}$ defined by $g_{ab}(\beta)$.

A basic idea in Perelman's approach[6] is to consider, along the solution $g_{ab}(\beta)$ of (11), a β-dependent mapping

$$f_\beta : \mathbb{R} \longrightarrow C^\infty(\Sigma_\beta, \mathbb{R}) \tag{13}$$
$$\beta \longmapsto f_\beta : \Sigma_\beta \to \mathbb{R},$$

where $C^\infty(\Sigma_\beta, \mathbb{R})$ denotes the space of smooth functions on Σ_β. In terms of f_β one constructs on Σ_β the β-dependent measure

$$d\varpi(\beta) \doteq (4\pi\tau(\beta))^{-\frac{3}{2}} e^{-f(\beta)} d\mu_{g(\beta)}, \tag{14}$$

where $\beta \longmapsto \tau(\beta) \in \mathbb{R}^+$ is a scale parameter chosen in such a way as to normalize $d\varpi(\beta)$ according to (*Perelman's coupling*)

$$\int_{\Sigma_\beta} d\varpi(\beta) = (4\pi\tau(\beta))^{-\frac{3}{2}} \int_{\Sigma_\beta} e^{-f(\beta)} d\mu_{g(\beta)} = 1. \tag{15}$$

It is easily verified that $\int_{\Sigma_\beta} d\varpi(\beta)$ is formally preserved along the Ricci flow (11), $\beta \in (0, \beta_*)$, *i.e.*,

$$\frac{\partial}{\partial \beta} \left[(4\pi\tau(\beta))^{-\frac{3}{2}} \int_{\Sigma_\beta} e^{-f(\beta)} d\mu_{g(\beta)} \right] = 0, \tag{16}$$

if the mapping f_β and the scale parameter $\tau(\beta)$ are evolved backward in time $\beta \in (\beta_*, 0)$ according to the coupled flows defined by

$$\begin{cases} \frac{\partial}{\partial\beta} f_\beta = -\Delta_{g(\beta)} f_\beta + \nabla^i f_\beta \nabla_i f_\beta - R(\beta) + \frac{3}{2}\tau(\beta)^{-1}, \ f(\beta_*) = f_* \\[2mm] \frac{\partial}{\partial\beta} \tau(\beta) = \frac{2}{3} \langle R(\beta) \rangle_{\Sigma_\beta(\beta)} \tau(\beta) - 1, \ \tau(\beta_*) = \tau_*, \end{cases} \tag{17}$$

where $\Delta_{g(\beta)}$ is the Laplacian with respect to the metric $g_{ab}(\beta)$, and f_*, τ_* are given (final) data, (backward β-evolution is required in order to have a well-posed parabolic initial value problem for (17)). (Note that the flows defined by (11) and (17) are based on the standard volume preserving Ricci flow and accordingly differ from those discussed by Perelman[6] which refer to the not-normalized flow $\frac{\partial}{\partial\eta}\widetilde{g}_{ab}(\eta) = -2\widetilde{R}_{ab}(\eta)$. Perelman's original flow and the one defined by (17) are related by the usual homotetic rescaling,[1,3,4] which maps the Ricci flow to the volume normalized flow[18]).

Let us remark that if $t \doteq \beta^* - \beta$, $\beta^* \in [0, T)$, then along a given volume-normalized Ricci flow $\beta \longmapsto g(\beta)$, $\beta \in [0, T)$, we have

$$\frac{\partial}{\partial t} d\Pi_t = [R(t) - \langle R(t)\rangle]\, d\Pi_t, \tag{18}$$

and Perelman's condition (17) yields the conjugated heat equation

$$\frac{\partial}{\partial t}\left[\frac{d\varpi(t)}{d\Pi_t}\right] = \Delta_{g(t)}\left[\frac{d\varpi(t)}{d\Pi_t}\right] - \frac{d\varpi(t)}{d\Pi_t}\left[R(t) - \langle R(t)\rangle\right], \tag{19}$$

where

$$\frac{d\varpi(t)}{d\Pi_t} = (4\pi\tau(t))^{-\frac{3}{2}}\, Vol[\Sigma_t]e^{-f(t)}. \tag{20}$$

Since $d\Pi_t$ is covariantly constant with respect to the Levi-Civita connection ∇ associated with $g(t)$

$$\nabla_i d\Pi_t = d\Pi_t\, g^{ab}(t)\nabla_i g_{ab}(t) = 0, \tag{21}$$

(this is equivalent to the familiar formula $\partial_i \ln \sqrt{g(t)} = \delta^c_a \Gamma^a_{ic}(t)$, where $\Gamma^a_{ic}(t)$ are the Christoffel symbols associated with $g_{ab}(t)$), we can exploit (18) and write (19) as the (non-uniformly parabolic) probability diffusion $(d\varpi(t))_{t\geq 0}$,

$$\begin{cases} \frac{\partial}{\partial t} d\varpi(t) = \Delta_{g(t)}\left(d\varpi(t)\right), \ t \doteq \beta^* - \beta \\[2mm] d\varpi(t = 0) = d\varpi_0. \end{cases} \tag{22}$$

Even if the dynamics (22) of the (backward) diffusion of $d\varpi(t)$ looks deceptively simple, what matters is the evolution equation (19) for the relative measure $\frac{d\varpi(t)}{d\Pi_t}$, and this shows that the dynamics of the (backward) diffusion of $\frac{d\varpi(t)}{d\Pi_t}$ along the Ricci flow is governed by the fluctuations in scalar curvature. This latter remark suggests that the natural modelling of $\frac{d\varpi(t)}{d\Pi_t}$ could be formulated within the framework of a Fokker-Planck dynamics describing a probability density undergoing diffusion in a potential field. To see if such a description is possible we need to characterize the tangent vector $\in T_{d\Pi_t} Prob(\Sigma)$ to the fiducial curve $t \to d\Pi_t$. Let us consider a generic value of the parameter t, say $t = s$. Since $\int_{\Sigma_t} \frac{\partial}{\partial t} d\Pi_t|_{t=s} = 0$, we have that $\frac{\partial}{\partial t} d\Pi_t|_{t=s} \in T_{d\Pi_t} Prob(\Sigma)$. According to (5), it is useful to parametrize $\frac{\partial}{\partial t} d\Pi_t|_{t=s}$ in terms of a scalar curvature-fluctuations potential Φ_s obtained as the solution of the elliptic equation

$$g^{ik}(s)\nabla_i\left(d\Pi_s\,\nabla_k\Phi_s\right) = -\left.\frac{\partial}{\partial t}d\Pi_t\right|_{t=s}, \tag{23}$$

where $\frac{\partial}{\partial t}d\Pi_t|_{t=s}$ is given by (18) for each given $t = s$, *i.e.*, (again exploiting the covariant constancy of $d\Pi_t$),

$$\Delta_{g(s)}\Phi_s = -\left(R(s) - \langle R(s)\rangle\right). \tag{24}$$

Formally, given a solution Φ_s of (24) and any (t-independent) smooth function with compact support $\zeta \in C_0^\infty(\Sigma,\mathbb{R})$, we can write

$$\frac{\partial}{\partial t}\int_{\Sigma_t}\zeta\,d\Pi_t = \int_{\Sigma_t}\nabla^i\zeta\,\nabla_i\Phi_t\,d\Pi_t = \langle\zeta,\Phi_t\rangle_{d\Pi_t}. \tag{25}$$

According to (3) the relation (25) identifies $d\Pi_t \longmapsto \Phi_t$ as the tangent vector to the curve $t \to d\Pi_t$ and defines the curvature (fluctuation) potential in which a probability density, evolving along a Ricci flow manifold, diffuses. Since $\int_\Sigma d\Pi_s\left(R(s) - \langle R(s)\rangle\right) = 0$, and

$$\int_{\Sigma_t}(R(t) - \langle R(t)\rangle_{\Sigma_t})\Phi_t\,d\Pi_t = \int_{\Sigma_t}|\nabla\,\Phi_t|^2\,d\Pi_t \tag{26}$$

we get that, as long as the (volume-normalized) Ricci flow exists, equation (24) admits a solution unique up to constants. The $L^2((\Sigma,d\Pi_s),\mathbb{R})$ norm of $(R(s) - \langle R(s)\rangle)$ is given by $\langle R^2(s)\rangle - \langle R(s)\rangle^2$, thus Φ_s is in the Sobolev space $H_2((\Sigma,d\Pi_s),\mathbb{R})$ if the mean square fluctuations in the scalar curvature are bounded. More generally, we know (see *e.g.*[4]) that if $\beta \mapsto g_{ab}(\beta)$ is a solution of the Ricci flow equation (for which the weak maximum principle holds), then bounds on the curvature (and its derivatives) of the initial metric induce a priori bounds on all derivatives $|\nabla^m R(x,s)|$ for a sufficiently

42

short time. Thus, for any given $t = s$ for which the Ricci flow is non-singular, we can assume that $(R(s) - \langle R(s)\rangle)$ is $C^\infty(\Sigma, \mathbb{R})$ and by elliptic regularity we get that $\Phi_s \in C^\infty(\Sigma, \mathbb{R})$.

3. Fokker-Planck diffusion along the Ricci flow

There is a useful consequence of the above parametrization of the curvature fluctuations which immediately shows why Fokker-Planck evolution is natural when we consider diffusion of a probability measure along the fiducial $d\Pi_t$.

Lemma 3.1. (^{18} *Lemma 2.4) For any probability measure $d\Omega \in Prob(\Sigma)$, absolutely continuous with respect to $d\Pi_t$, the following identity holds along the the volume-normalized Ricci flow*

$$\frac{\partial}{\partial t} S\left[d\Omega \parallel d\Pi_t\right] = -I\left[d\Omega \parallel d\Pi_t\right] + \tag{27}$$

$$+ \int_{\Sigma_t} \ln \frac{d\Omega}{d\Pi_t} \left[\frac{\partial}{\partial t} d\Omega - \Delta_{g(t)}\, d\Omega + \nabla^i\left(d\Omega \nabla_i \Phi_t\right)\right],$$

where

$$S\left[d\Omega \parallel d\Pi_t\right] \doteq \int_{\Sigma_t} d\Omega \ln \frac{d\Omega}{d\Pi_t}, \tag{28}$$

and

$$I\left[d\Omega \parallel d\Pi_t\right] \doteq \int_{\Sigma_t} d\Omega\; \nabla^i \ln \frac{d\Omega}{d\Pi_t}\; \nabla_i \ln \frac{d\Omega}{d\Pi_t} \tag{29}$$

respectively denote the relative entropy of $d\Omega$ with respect to $d\Pi_t$ and the associated entropy generating functional. Moreover, one computes

$$\frac{\partial}{\partial t} I\left[d\Omega \parallel d\Pi_t\right] = \tag{30}$$

$$= \int_{\Sigma_t} d\Omega \left[\left|\nabla \ln \frac{d\Omega}{d\Pi_t}\right|^2 + 2\nabla^i \ln \frac{d\Omega}{d\Pi_t} \nabla_i\right] \left\{\frac{\partial}{\partial t} \ln \frac{d\Omega}{d\Pi_t} - \right.$$

$$\left. - \Delta_{g(t)} \ln \frac{d\Omega}{d\Pi_t} + \left|\nabla \ln \frac{d\Omega}{d\Pi_t}\right|^2 - \nabla^i \Phi_t \nabla_i \ln \frac{d\Omega}{d\Pi_t}\right\} +$$

$$+ \int_{\Sigma_t} d\Omega \left[-2R^{ik}(t) + \frac{2}{3} \langle R(t)\rangle_{\Sigma_t}\, g^{ik}(t)\right] \nabla_i \ln \frac{d\Omega}{d\Pi_t}\, \nabla_k \ln \frac{d\Omega}{d\Pi_t} -$$

$$- 2\int_{\Sigma_t} d\Omega \left[R^{ik}(t) + (Hess\, \Phi_t)^{ik}\right] \nabla_i \ln \frac{d\Omega}{d\Pi_t}\, \nabla_k \ln \frac{d\Omega}{d\Pi_t} -$$

$$- 2\int_{\Sigma_t} d\Omega \left|Hess\left(\ln \frac{d\Omega}{d\Pi_t}\right)\right|^2.$$

From this Lemma, we immediately get the following

Theorem 3.1. (^{18} *Th. 2.5*) *The Fokker-Planck diffusion* $(d\Omega_t)_{t\geq 0}$ *generated, along the volume-preserving Ricci flow, by*

$$\frac{\partial}{\partial t}\left(d\Omega_t\right) = \Delta_{g(t)}d\Omega_t - \nabla^i\left(d\Omega_t\nabla_i\Phi_t\right), \tag{31}$$

is the gradient flow of the relative entropy functional $S\left[d\Omega_t \parallel d\Pi_t\right]$, *i.e.,*

$$\frac{\partial}{\partial t}\int_{\Sigma_t} d\Omega_t \, \ln\frac{d\Omega_t}{d\Pi_t} = -\int_{\Sigma_t} d\,\Omega_t\nabla_i \ln\left(\frac{d\Omega_t}{d\Pi_t}\right)\nabla^i \ln\left(\frac{d\Omega_t}{d\Pi_t}\right). \tag{32}$$

Moreover, the associated entropy production functional $I\left[d\Omega_t \parallel d\Pi_t\right]$ *satisfies the differential inequality*

$$\frac{\partial}{\partial t}I\left[d\Omega_t \parallel d\Pi_t\right] \leq -2\,K_t\,I\left[d\Omega_t \parallel d\Pi_t\right] \tag{33}$$

where $K_t \in \mathbb{R}$ *is the (t-dependent) lower bound of the Ricci curvature.*

Note that we can rewrite (33) as

$$\frac{\partial^2}{\partial t^2}S[d\Omega_t \parallel d\Pi_t] \geq 2K_tI[d\Omega_t \parallel d\Pi_t] \tag{34}$$

which shows that the lower bound of the Ricci curvature, besides controlling the rate of production of entropy along the Fokker-Planck diffusing $d\Omega_t$, is related to the convexity properties of $S[d\Omega_t \parallel d\Pi_t]$. Moreover, according to (10),

$$S[d\Omega_t \parallel d\Pi_t] \geq \frac{1}{2}\left[\frac{D_s^W(d\Omega_t, d\Pi_t)}{diam(\Sigma, g(t))}\right]^{2s}, \tag{35}$$

thus, the above result implies also that the Wasserstein distance $D_s^W(d\Omega_t, d\Pi_t)$ is weakly monotonically decreasing with t, and $d\Omega_t$ approaches the reference Ricci measure $d\Pi_t$ minimizing the relative entropy functional $S[d\Omega_t \parallel d\Pi_t]$.

In its simplest form, the rate of convergence of a solution of the Fokker-Planck equation

$$\begin{cases} \frac{\partial}{\partial t}\frac{d\Omega_t}{d\Pi_t} = \Delta_{g(t)}\frac{d\Omega_t}{d\Pi_t} - \nabla^i\Phi_t\nabla_i\frac{d\Omega_t}{d\Pi_t}, \\[2mm] \frac{d\Omega_t}{d\Pi_t}\big|_{t=0} = \frac{d\Omega_0}{d\Pi_0}, \end{cases} \tag{36}$$

to the stationary state $d\Pi_t$ is governed by the following result which improves (35) and establishes the relevant asymptotics for the Fokker-Planck diffusion

Lemma 3.2. *(^{18} Lemma 2.6) Let $\beta \to g_{ab}(\beta)$, $\beta \in [0, \infty)$, a given Ricci flow metric starting on a manifold (Σ, g) of positive Ricci curvature. Then the entropy functional $S\left[d\Omega_t \parallel d\Pi_t\right]$ decreases exponentially fast according to*

$$S\left[d\Omega_t \parallel d\Pi_t\right] \leq S\left[d\Omega_0 \parallel d\Pi_0\right] e^{-\frac{2}{3}\lambda_{\inf} \, t}, \tag{37}$$

with $\frac{1}{3}\lambda_{\inf} \doteq \inf_{\beta \geq 0} \{K_\beta > 0 : Ric(\beta) \geq K_\beta \, g(\beta)\}$. Moreover, the Talagrand inequality

$$\left[D_2^W(d\Omega_t, d\Pi_t)\right]^2 \leq \frac{6}{\lambda_{\inf}} \, S\left[d\Omega_t \parallel d\Pi_t\right] \tag{38}$$

holds, and $S\left[d\Omega_t \parallel d\Pi_t\right]$ is a convex function along the Ricci flow. In particular, if we introduce the adimensional variable $\eta \doteq \frac{1}{3}\lambda_{inf} \, t$ then

$$\frac{\partial^2}{\partial \eta^2} S\left[d\Omega_t \parallel d\Pi_t\right] \geq \frac{2}{3}\lambda_{inf} \left[D_2^W(d\Omega_t, d\Pi_t)\right]^2, \tag{39}$$

which is equivalent to Sturm's K-convexity^{22} of $S\left[d\Omega_t \parallel d\Pi_t\right]$, (for $K = \frac{1}{3}\lambda_{inf}$).

It is worthwhile to note that geodesic convexity for relative entropy was conjectured to hold for Riemannian manifolds with non-negative Ricci curvature by F. Otto and C. Villani.[25] In the case $K = 0$, the conjecture has been proven in[30] whereas in the general case (for any $K \in \mathbb{R}$) in.[22] The fact that (39) holds is a strong indications that Ricci flow theory can be extended to metric spaces more general than smooth Riemannian manifolds. This is a subject of considerable interest since there is no reason why Thurston's geometrization program should not apply to three-dimensional metric spaces of a rather general nature, from Piecewise Linear (PL) metrized manifolds (Regge manifolds[31]) to the more general class of length spaces. It is important to stress that for PL manifolds there is available a sophisticated approach to the combinatorial counterpart of the Ricci flow (see the remarkable[32] and[33,34]) which shows that a direct extension of geometric flow theory to non-smooth manifolds is far from being trivial. It is thus promising that optimal trasportation methods, which are naturally extended to general metric spaces,[35] can be made available in such a setting.

Aknowledgements

The author would like to thank the organizers for a very stimulating and enjoyable meeting.

Research supported in part by PRIN Grant #2004012375 − 002.

References

1. R. S. Hamilton, *Three-manifolds with positive Ricci curvature*, J. Diff. Geom. **17**, 255-306 (1982).
2. M. Carfora, A. Marzuoli, *Model geometries in the space of Riemannian structures and Hamilton's flow*, Class. Quantum Grav. 5 (1988) 659-693.
3. T. Aubin, *Some nonlinear problems in Riemannian Geometry*, Springer Verlag (1998).
4. B. Chow, D. Knopf, *The Ricci Flow: An Introduction*, Math. Surveys and Monographs **110**, (2004) Am. Math. Soc.
5. R. S. Hamilton, *The formation of singularities in the Ricci flow*, Surveys in *Differential Geometry Vol 2*, International Press, (1995) 7–136.
6. G. Perelman *The entropy formula for the Ricci flow and its geometric applications* math.DG/0211159.
7. T. Buchert and M. Carfora, *Regional averaging and scaling in relativistic cosmology* Class. Quant. Grav. **19**, (2002) 6109-6145.
8. T. Buchert and M. Carfora, *Cosmological parameters are dressed*, Phys. Rev. Lett. **90**, (2003) 31101-1-4.
9. I. Bakas, *Geometric flows and (some of) their physical applications*, AvH conference Advances in Physics and Astrophysics of the 21st Century, 6-11 September 2005, Varna, Bulgaria, hep-th/0511057.
10. D. H. Friedan, *Nonlinear models in* $2 + \varepsilon$ *dimensions*, Ann. Physics **163** (1985), no. 2, 318–419.
11. T Oliynyk, V Suneeta, E Woolgar *A Gradient Flow for Worldsheet Nonlinear Sigma Models*, Nucl.Phys. B739 (2006) 441-458, hep-th/0510239.
12. G. Perelman *Ricci flow with surgery on Three-Manifolds* math.DG/0303109.
13. G. Perelman *Finite extinction time for the solutions to the Ricci flow on certain three-manifolds* math.DG/0307245.
14. M. Feldman, T.Ilmanen, Lei Ni, *Entropy and reduced distance for Ricci expanders*, J. Geom. Anal. **15** (2005), no. 1, 49–62.
15. Lei Ni, *The entropy formula for linear heat equation*, J. Geom. Anal. **14** (2004), no. 1, 87–100.
16. Lei Ni, Addenda to: *The entropy formula for linear heat equation* J. Geom. Anal. **14** (2004), no. 2, 369–374.
17. L. Gross, *Logarithmic Sobolev inequalities and contractivity properties of semigroups*, Dirichlet forms. Lectures given at the First C.I.M.E. Session held in Varenna, June 8–19, 1992. Edited by G. Dell'Antonio and U. Mosco. Lecture Notes in Mathematics, 1563. Springer-Verlag, Berlin, 1993.
18. M. Carfora, *Fokker-Planck dynamics and entropies for the normalized Ricci flow*, arXiv:math.DG/0507309 v2.

19. K.T. Sturm, *On the geometry of metric measure spaces I, II* preprints (2004,2005).

20. J. Lott, C. Villani, *Ricci curvature for metric-measure spaces via optimal. transport*, preprint http://www.arxiv.org/abs/math.DG/0412127 (2004).

21. John Lott, Cedric Villani, *Weak curvature conditions and Poincare inequalities*, preprint http://www.arxiv.org/abs/math.DG/0506481 (2005).

22. M.-K. von Renesse, K. T. Sturm, *Transport inequalities, gradient estimates, entropy and Ricci curvature*, Comm. Pure Appl. math. **58** (2005) 1-18.

23. J-D. Deuschel and D. Strook, *Large Deviations*, Academic Press (1989).

24. F. Otto, *The geometry of dissipative evolution equations: the porous medium equation*, Comm. Partial Diff. Equations **23** (2001), 101-174.

25. F. Otto and C. Villani, *Generalization of an inequality by Talagrand, and links with the logarithmic Sobolev inequality*, J. Funct. Anal., **173(2)** (2000), 361-400.

26. G. Toscani, C. Villani *On the trend to equilibrium for some dissipative systems with slowly increasing a priori bounds*, J. Statist. Phys. **98**, (2000), 1279-1309.

27. C. Villani, *Topics in Optimal Transportation*, Am. Math. Soc. Providence, RI, (2003).

28. S.-I. Ohta, *On measure contraction property of metric measure spaces* Preprint (http://www.math.kyoto-u.ac.jp/ sohta/) (2005).

29. F. Bolley, C. Villani, *Weighted Csiszar-Kullback-Pinsker inequalities and applications to transportation inequalities*, Ann. Fac. Sci. Toulose Math. (to appear).

30. D. Cordero-Erausquin, R.J. McCann, M. Schmuckenschlaegerger, *A Riemannian interpolation inequality a' la Borell, Brascamp and Lieb*, Invent. Math. 146 (2001), 219-257.

31. T. Regge, *General relativity without coordinates*, Nuovo Cimento (10) 19 (1961), 558-571.

32. D. Glickenstein, *A combinatorial Yamabe flow in three dimensions*, Topology 44 (2005) 791-808, math.MG/0506182 (see also by the same author math.MG/0211195)

33. B. Chow and F. Luo, *Combinatorial Ricci flows on surfaces*, J. Differential Geom. 63 (2003), 97-129.

34. F. Luo, *Combinatorial Yamabe flow on surfaces*, arXiv:math.GT/0306167, 2003.

35. L. Ambrosio, N. Gigli, G. Savaré, *Gradient flows in metric spaces and in the space of probability measures*, Lectures in Mathematics ETH Zürich, Birkhäuser Verlag, Basel (2005) viii+333

EXACT SOLUTIONS OF A
REACTION DIFFUSION EQUATION

M. CARINI

Department of Mathematics, University of Calabria
v. P. Bucci, cubo 30B,
Arcavacata di Rende, Cosenza, Italy
E-mail: carini@unical.it

N. MANGANARO

Department of Mathematics, University of Messina,
Contrada Papardo, Salita Sperone 31,
98166 Messina, Italy
E-mail: nat@mat520.unime.it

Dedicated to A. Greco in occasion of his 65th birthday

A model equation which describes reaction diffusion processes is considered. We classify the possible forms of the material response functions involved in the governing equation under interest allowing generalized separable exact solutions to hold. Exact solutions of a nonlinear second order reaction diffusion equation are calculated.

1. Introduction

In this paper we consider an equation of the form

$$u_t = \partial_{xx}\left\{F\left(x,u\right)\right\} + G\left(x,u\right) \tag{1}$$

which models reaction diffusion processes. In (1) $u(x,t)$ is the field variable, x and t are, respectively, space and time coordinates, $F(x,u)$ and $G(x,u)$ are material response functions. Moreover subscripts with respect to x and t mean for partial derivatives.

48

The model (1) generalizes many equations well known in the litterature. For instance, when $F = A(u)$ and $G = C(u)$, equation (1) specializes to

$$u_t = \partial_x \left\{ A'(u) u_x \right\} + C(u) \tag{2}$$

where the prime indicates ordinary differentiation. Equation (2) describes many processes of interest in different fields as in microwave heating (see Hill and Smith[1] and Hill and Pincombe[2]); in hot magnetized fusion plasmas (see Le Roux and Wilhelmsson[3]), in biophysics, solid state physics, reactor design (Murray,[4] Scott,[5] Meyrs et al.,[6] Aris[7]).

When $C(u) = 0$, equation (2) gives the very well known nonlinear heat equation, while when $A(u) = u$ we get

$$u_t = u_{xx} + C(u). \tag{3}$$

In particular, if $C(u) = au - bu^2$, from (3) we deduce the Fisher's equation[8] which is of great interest in mathematical biology and in mathematical ecology describing time and space evolution of the density of individuals as bacteria; if $C(u) = u(1 - u^2)$, the Newell–Whitehead equation[9] is found; if $C(u) = u(1-u)(u-u_0)$, where u_0 is a constant, the Fitzhugh–Nagumo equation, which describes the transmission of nerve impulses (see Fitzhugh[10] and Nagumo et alt.[11]) and different processes of interest in population genetics (see Aronson and Weinberger[12,13]), is obtained; if $C(u) = u^2(1 - u)$, the Huxley equation is recovered.

As far as the research of exact solutions to (2) is concerned, a large body of litterature exists. Lie[14] calculates the classical symmetries of the linear heat equation ($A(u) = u$ and $C(u) = 0$), while Bluman and Cole[15] determine its non–classical symmetries. Ovsiannikov[16] describes the Lie symmetries of the nonlinear heat equation ($C(u) = 0$). Dorodnitsyn[17] characterizes Lie symmetries of (2), while conditional symmetries are studied by Serov,[18] Arrigo et al.,[19] Clarkson and Mansfield,[20] Fushchych et al.[21] A large body of references concerning classes of exact solutions to (2) can be found in Arrigo and Hill.[22]

Several exact solutions are also known for equation (3). Classical and non–classical symmetries are calculated by Clarkson and Mansfield;[20] conditional symmetries are determined by Fushchich and Serov;[23] non–local symmetries are characterized by Tsyra and Fushchych.[24]

In the case of the Fisher's equation ($C(u) = au - bu^2$), travelling waves, similarity solutions and exact solutions in terms of Jacobian elliptic functions and using Painleve expansion method are obtained by Ablowitz et al.,[25] Guo et al.,[26] Kaliappan et al.,[27] Herrera et al.,[28] Bindu et al.,[29] Kenre,[30,31] Abramson et al.,[32] Brazhnik et al.[33]

In the case of Newell–Whitehead equation $(C(u) = u(1 - u^2))$, Cariello and Tabor[34,35] find exact solutions using Painleve expansion, while several exact solutions of the Fitzhugh–Nagumo equation $(C(u) = u(1 - u)(u - u_0))$ are determined using different approaches (see Fife,[36] Veling,[37] Vorob'ev,[39] Kawahara,[40] Nucci[41]). Exact solutions of the Huxley equation $(C(u) = u^2(1 - u))$ are obtained by Chen and Guo[42] and by Clarkson and Mansfield.[20]

Finally differential invariants of equivalence transformations for the equation

$$u_t = h(x, u)u_{xx} + c\,(x, u, u_x)\,. \tag{4}$$

are calculated by Ibragimov and Sophocleous.[44]

Looking for exact solutions of model equation like (1) is of great interest not only from a theoretical point of view but also for the applications. In fact, exact solutions could tell us a lot about the dynamics of the model under interest as well as about many interesting behaviours as "wating–time" or "blow up". Furthermore they can also be used to test numerical procedures useful to integrate the equations at hand.

When in the governing model we have some unspecified functions, another problem is to select the forms of the material response functions (Model Constitutive Laws) involved therein in order that the governing model admits the class of solutions we are interested in.

One of the most simple and nevertheless interesting class of exact solutions is the additive separable form $(u = \alpha(x) + \beta(t))$ and the product separable solution $(u = \alpha(x)\beta(t))$. In the paper by Pucci and Saccomandi[45] the connections between such a classes of solutions and group theoretical approaches are well clarified.

The aim of this paper is to look for exact solutions of (1) generalizing the additive and product forms. In particular we classify the forms of the material response functions $F(x, u)$ and $G(x, u)$ allowing the solutions we are interested in be admitted by (1).

The paper is organized as follows. In section 2 we develop a general procedure based on the theory of differential constraints (see Ianenko[46]) classifying model (1). In section 3 some exact solutions of (2) and (3) are presented. Section 4 is devoted to some conclusions and final remarks.

2. Reduction Procedure

In order to classify the model equation (1) with respect generalized additive and product separable solutions, let us notice that if a solution in the form

50

$u = H(x) + \phi(t)$ or $u = H(x)P(t)$ satisfies the governing equation, then the conditions $u_x = H'(x)$ or $u_x = H'(x)P(t)$ are compatible with the equation at hand. Thus, here, we look for exact solutions of (1) which satisfy also a differential constraint of the form

$$u_x = H'(x) + P(t) \tag{5}$$

or

$$u_x = H'(x)P(t). \tag{6}$$

By appending the relation (5) or (6) to the equation (1), an overdetermined system is obtained and the differential compatibility has to be required. Once the resulting compatibility condition has been satisfied, from (5) or (6) exact solutions of (1) under the form

$$u = H(x) + xP(t) + \phi(t) \tag{7}$$

or

$$u = H(x)P(t) + \phi(t) \tag{8}$$

will be determined.

As first case we append to the model equation (1) the differential constraint (5). The resulting compatibility condition writes:

$$\begin{aligned}
&3F_{xxu}H_x + F_uH_{xxx} + 3F_{xu}H_{xx} + F_{xxx} + G_x - P_t+ \\
&+F_{uuu}H_x^3 + F_{uuu}P^3 + G_uH_x + G_uP + 3F_{xxu}P + 3F_{xuu}H_x^2+ \\
&+3F_{xuu}P^2 + 3F_{uuu}H_xP^2 + 6F_{xuu}H_xP + 3F_{uuu}H_x^2P+ \\
&+3F_{uu}H_{xx}H_x + 3F_{uu}H_{xx}P = 0.
\end{aligned} \tag{9}$$

Solving relation (9), we characterize the functions $H(x)$ and $P(t)$ involved in (7) as well as the forms of the material response functions $F(x, u)$ and $G(x, u)$ allowing the solution (7) to exist.

If $P'(t) \neq 0$, after some algebra, we solve (9) and we can prove that the equation (1) admits an exact solution of the form (7) if $F(x, u)$ and $G(x, u)$ specialize to

$$F(x, u) = a_3u^3 + a_2(x)u^2 + a_1(x)u \tag{10}$$

$$G(x, u) = (k_0 - 6\alpha_2)\,u + \beta_0(x) \tag{11}$$

where

$$a_2(x) = -3a_3 H(x) + c_1 x + c_0; \tag{12}$$

$$a_1(x) = 3a_3 H^2(x) - 2(c_1 x + c_0) H(x) + \alpha_2 x^2 + \alpha_1 x + \alpha_0 \tag{13}$$

$$\beta_0(x) = 2c_1 \left(xHH_{xx} + 2HH_x + xH_x^2\right) - \alpha_2 \left(x^2 H_{xx} - 4H + 4xH_x\right) +$$
$$-\alpha_1 \left(xH_{xx} + 2H_x\right) - 3a_3 \left(H^2 H_{xx} + 2HH_x^2\right) + 2c_0 \left(H_x^2 + HH_{xx}\right) +$$
$$-\alpha_0 H_{xx} - k_0 H + k_1 x + \beta_0 \tag{14}$$

while a_3, c_1, c_0, α_2, α_1, α_0, k_0, k_1 and β_0 are constants. Moreover the functions $P(t)$ and $\phi(t)$ involved in (7) can be calculated by solving the equations

$$P'(t) = 6a_3 P^3 - 6c_1 P^2 + k_0 P + k_1; \tag{15}$$

$$\phi'(t) + \left(4\alpha_2 - k_0 - 4c_1 P + 6a_3 P^2\right)\phi(t) = 2\alpha_1 P + 2c_0 P^2 + \beta_0. \tag{16}$$

while $H(x)$ is unspecified.

If $P = 0$, condition (9) gives a relation to which $F(x,u)$ and $G(x,u)$ must satisfy in order that the classical additive separable solution (7) is admitted by (1).

As second case we append to the equation (1) the differential constraint (6), whereupon the resulting compatibility condition is

$$3F_{xxu}H_x P + 3F_{xuu}H_x^2 P^2 + F_{uuu}H_x^3 P^3 + 3F_{uu}H_x H_{xx}P^2 + H_x PG_u +$$
$$-P_t + F_{xxx} + 3F_{xu}H_{xx}P + F_u H_{xxx}P + G_x P - H_x P_t = 0. \tag{17}$$

After some algebra, the solution of (17) leads to the following two cases.

i)

$$F(x,u) = a_3 u^3 + a_2(x)u^2 + a_1(x)u \tag{18}$$

$$G(x,u) = (k1 - 6\alpha_2)u + c_1 k_0 x + \beta_0 \tag{19}$$

$$H(x) = c_1 x + c_0 \tag{20}$$

where

$$a_3 = \frac{k_3}{6c_1^2}; \quad a_2(x) = \frac{k_2}{6c_1}x + \epsilon_0; \tag{21}$$

$$a_1(x) = \alpha_2 x^2 + \alpha_1 x + \alpha_0 \tag{22}$$

while the functions $P(t)$ and $\phi(t)$ involved in (8) must satisfy the equations

$$P'(t) = k_3 P^3 + k_2 P^2 + k_1 P + k_0; \tag{23}$$

$$\phi'(t) + \left(4\alpha_2 - k_1 - \frac{2}{3}k_2 P - k_3 P^2\right)\phi(t) =$$
$$= 2\alpha_1 c_1 P + 2\epsilon_0 c_1^2 P^2 + \beta_0. \tag{24}$$

52

In (18)–(24) k_1, α_2, c_1, k_0, c_0, k_3, k_2, α_1, α_0 are arbitrary constants.

   ii)

$$F(x, u) = a_2(x)u^2 + a_1(x)u \tag{25}$$

$$G(x, u) = F_{xx} + k_0 H(x) + \beta_1 u + \beta_0 \tag{26}$$

where

$$a_2(x) = \frac{k_2 x + \alpha_2}{6 H_x} \tag{27}$$

while the functions $a_1(x)$ and $H(x)$ are determined by solving the equations

$$(\beta_1 - k_1) H(x) + 2H'(x)a_1'(x) + H''a_1(x) = \epsilon_0; \tag{28}$$

$$(k_2 x + \alpha_2) \frac{d}{dx} \left(\ln \left(\frac{H'}{H''} \right) \right) = k_2 \tag{29}$$

and k_2, α_2, β_1, β_0, ϵ_0 are arbitrary constants. Moreover the functions $P(t)$ and $\phi(t)$ involved in (8) satisfy the equations

$$P'(t) = k_2 P^2 + k_1 P + k_0, \tag{30}$$

$$\phi'(t) + (\mu_0 P(t) - \beta_1) \phi(t) = \beta_0 + \epsilon_0 P + \mu_1 P^2 \tag{31}$$

where k_0, k_1, μ_0 and μ_1 are constants.

Summarizing we proved the following theorem:

Theorem 2.1. *In order that the model equation (1) admits generalized additive or product separable exact solutions of the form (7) or (8), the material response functions $F(x, u)$ and $G(x, u)$ must adopt one of the expressions (10), (11) or (18), (19) or (25), (26).*

3. Exact Solutions

The aim of this section is to specialize some of the results obtained in the previous section in order to characterize exact solutions of the equation (2). Thus, here and in what follows, we assume

$$F = F(u); \qquad G = G(u). \tag{32}$$

First, let us look for solutions of the equation (1) supplemented by (32) under the form (8). Taking relations (18)–(24) into account, as first case we get.

   i)

$$F(x, u) = a_3 u^3 + a_2 u^2 + a_1 u \tag{33}$$

$$G(x, u) = \beta_1 u + \beta_0 \tag{34}$$

where a_3, a_2, a_1, β_1, β_0 are constants. By integrating (23) and (24), the corresponding exact solutions are:

i_1) if $a_3 \neq 0$ and $\beta_1 \neq 0$ (we assume without loss of generality $\beta_0 = 0$) we obtain

$$u(x,t) = c_1 x \sqrt{\frac{\beta_1}{e^{-2\beta_1 t} - \mu}} + \phi(t), \qquad \mu = 6a_3 c_1^2 \tag{35}$$

where, if $\mu = \mu_0^2$, then

$$\phi(t) = \frac{1}{\sqrt{e^{-2\beta_1 t} - \mu_0^2}} \left(-\frac{a_2 \mu_0}{3a_3} \arctan\left(\frac{\sqrt{e^{-2\beta_1 t} - \mu_0^2}}{\mu_0} \right) + \phi_0 \right); \tag{36}$$

if $\mu = -\mu_0^2$, then

$$\phi(t) = \frac{1}{\sqrt{e^{-2\beta_1 t} + \mu_0^2}} \left(\frac{a_2 \mu_0}{6a_3} \ln\left(\frac{\sqrt{e^{-2\beta_1 t} + \mu_0^2} - \mu_0}{\sqrt{e^{-2\beta_1 t} + \mu_0^2} + \mu_0} \right) + \phi_0 \right); \tag{37}$$

i_2) if $a_3 \neq 0$ and $\beta_1 = 0$, we deduce

$$u(x,t) = \frac{x}{\sqrt{12a_3 (t - t_0)}} + \phi(t), \tag{38}$$

where

$$\phi(t) = \frac{2}{3} \beta_0 (t - t_0) + \frac{\phi_0}{\sqrt{|t - t_0|}} - \frac{a_2}{3a_3}; \tag{39}$$

i_3) if $a_3 = 0$ and $\beta_1 \neq 0$, we determine

$$u(x,t) = c_1 x e^{\beta_1 t} + \phi(t), \tag{40}$$

where

$$\phi(t) = e^{\beta_1 t} \left(\frac{2a_2 c_1^2}{\beta_1} e^{\beta_1 t} + \phi_0 \right); \tag{41}$$

i_4) if $a_3 = \beta_1 = 0$ the trivial solution

$$u(x,t) = c_1 x + \left(\beta_0 + 2a_2 c_1^2 \right) t \tag{42}$$

is obtained. Moreover in (36), (37), (39) and (41) ϕ_0 denotes an arbitrary constant.

As second case we have.

ii)

$$F(x,u) = a_2 u^2 + a_1 u \tag{43}$$
$$G(x,u) = \beta_1 u + \beta_0 \tag{44}$$

where a_2, a_1, β_1, β_0 are constants. Owing relations (27)–(31), the corresponding exact solutions are:

$ii_1)$ if $\beta_1 \neq 0$, we get

$$u(x,t) = \frac{c_2 \beta_1 x^2}{e^{-\beta_1 t} + \nu} + \phi(t), \qquad \nu = -12 a_2 c_2;$$
(45)

where

$$\phi(t) = e^{\frac{2}{3}\beta_1 t} \left(e^{-\beta_1 t} + \nu\right)^{1/3} \left\{ 2 a_2 c_2 \left[\ln\left(\left(\nu e^{\beta_1 t} + 1\right)^{1/3} - 1\right) + \right.\right.$$
$$\left. - \tfrac{1}{2}\ln\left(\left(\nu e^{\beta_1 t} + 1\right)^{2/3} + \left(\nu e^{\beta_1 t} + 1\right)^{1/3} + 1\right)\right]$$
$$\left. - \sqrt{3}\arctan\left\{\tfrac{\sqrt{3}}{3}\left(2\left(\nu e^{\beta_1 t} + 1\right)^{1/3} + 1\right)\right\} + \phi_0\right\};$$
(46)

$ii_2)$ if $\beta_1 = 0$, we deduce

$$u(x,t) = \frac{x^2}{12 a_2 \left(t_0 - t\right)} + \phi(t),$$
(47)

where

$$\phi(t) = \phi_0 \left(t - t_0\right)^{-\frac{1}{3}} + \frac{3}{4}\beta_0 \left(t - t_0\right) - \frac{a_1}{2 a_2}.$$
(48)

In (46) and (48) ϕ_0 is an arbitrary constant.

As third case we consider

$iii)$

$$F(x,u) = a_1 u, \qquad G(x,u) = \beta_1 u + \beta_0$$
(49)

where a_1, β_1, β_0 are constants. Taking relations (27)–(31) into account, the corresponding exact solution writes

$$u(x,t) = e^{kt} H(x) + \phi(t),$$
(50)

with k an arbitray constant, while $H(x)$ adopts one of the following forms: if $\frac{\beta_1 - k}{a_1} = -\mu^2$, then

$$H(x) = \gamma_1 e^{-\mu x} + \gamma_2 e^{\mu x} - \frac{c_0}{\mu^2};$$
(51)

if $\frac{\beta_1 - k}{a_1} = \mu^2$, then

$$H(x) = \gamma_1 \cos\left(\mu x\right) + \gamma_2 \sin\left(\mu x\right) + \frac{c_0}{\mu^2};$$
(52)

if $\beta_1 = k$, then

$$H(x) = \frac{c_0}{2} x^2 + \gamma_1 x + \gamma_2;$$
(53)

where γ_1, γ_2 and c_0 are constants. Furthermore the function $\phi(t)$ involved in (42) must adopt one of the expressions:
if $\beta_1 \neq k$ and $\beta_1 \neq 0$, then

$$\phi(t) = -\frac{c_0 a_1}{\beta_1 - k} e^{kt} + \phi_0 e^{\beta_1 t}; \tag{54}$$

if $\beta_1 = k$ and $\beta_1 \neq 0$, then

$$\phi(t) = (\phi_0 - c_0 a_1 t)\, e^{\beta_1 t}; \tag{55}$$

if $\beta_1 = 0$ and $k \neq 0$, then

$$\phi(t) = \beta_0 t - \frac{c_0 a_1}{k} e^{kt} + \phi_0; \tag{56}$$

if $\beta_1 = k = 0$, then

$$\phi(t) = (\beta_0 - c_0 a_1)\, t + \phi_0; \tag{57}$$

where ϕ_0 is an arbitrary constant.

As final case we consider exact solutions in the form (7). By specializing the results obtained in (10)–(16), we find that the material response functions $F(u)$ and $G(u)$ must adopt the form (49), while the functions $H(x)$, $P(t)$ and $\phi(t)$ involved in (7) are given as it follows:
if $\beta_1 \neq 0$, then

$$P(t) = -\frac{c_0}{\beta_1} + p_0 e^{\beta_1 t} \tag{58}$$

$$\phi(t) = \phi_0 e^{\beta_1 t} \tag{59}$$

while

$$H(x) = m_0 e^{-\mu x} + m_1 e^{\mu x} + \frac{c_0}{\beta_1} x \quad \text{if} \quad \frac{\beta_1}{a_1} = -\mu^2 \tag{60}$$

$$H(x) = m_0 \cos(\mu x) + m_1 \sin(\mu x) + \frac{c_0}{\beta_1} x \quad \text{if} \quad \frac{\beta_1}{a_1} = \mu^2; \tag{61}$$

if $\beta_1 = 0$, then

$$P(t) = c_0 t + p_0 \tag{62}$$

$$\phi(t) = (\beta_0 + c_1 a_1)\, t \tag{63}$$

$$H(x) = \frac{c_0}{6 a_1} x^3 + \frac{c_1}{2 a_1} x^2 + m_0 x + m_1. \tag{64}$$

In (58)–(64) c_0, c_1, p_0, ϕ_0, m_0 and m_1 are constants.

4. Conclusions and Final Remarks

In this paper the compatibility of the model equation (1) with classes of generalized separable exact solutions of the form (7) or (8) is studied. We prove that the material response functions $F(x, u)$ and $G(x, u)$ involved in (1) must assume the forms $F(x, u) = a_3(x)u^3 + a_2(x)u^2 + a_1(x)$, $G(x, u) = \beta_2(x)u^2 + \beta_1(x)u + \beta_0(x)$ in order that the governing model equation (1) admits exact solutions belonging to the classes (7) or (8). Generalized separable exact solutions of the nonlinear reaction diffusion equation (2) are calculated.

It should be of a certain interest to study special behaviours as "blow up" of the solutions at hand as well as to develop a stability analysis. These topics will be considered in a future paper.

Acknowledgments

This work was partially supported by MURST, Progetto di Cofinanziamento 2005 "Nonlinear Propagation and Stability in Thermodynamical Processes of Continuous Media" and by Fondi del Programma di Ricerca Ordinario 2003 (PRA 2003) "Metodologie di Riduzione e Risoluzione di Problemi Iniziali e/o al Contorno per Equazioni di Evoluzione" of University of Messina.

References

1. J. M. Hill and N. F. Smith, *Math. Engrg. Ind.*, **2**, 267–278, (1990).
2. J. M. Hill and H.Pincombe, *J. Austral. Math. Soc. Ser. B*, **33**, 290–320, (1992).
3. M. N. Le Roux and H. Wilhelmsson, *Phys. Scripta*, **40**, 674–681, (1989).
4. J.D. Murray, *Mathematical Biology*, Springer–Verlag, Berlin (1989).
5. A. C. Scott, *Rev. Mod. Phys.*, **47**, 487–533, (1975).
6. A. K. Meyrs–Beaghton and D.D. Vedensky, *Phys. Rev. B*, **42**, 5544–5554, (1990).
7. R. Aris, *The Mathematical Theory of Diffusion and Reaction in Permeable Catalysts*, **I–II**, Clarendon, Oxford (1989).
8. R. A. Fisher, *Ann. of Eugenics*, **7**, 355–369, (1937).
9. A. C. Newell and J. A. Whitehead, *J. Fluid Mech.*, **38**, 279–303, (1969).
10. R. Fitzhugh, *Biophysical J.*, **1**, 445–466, (1937).
11. J. S.Nagumo, S. Arimoto and S. Yoshizawa, *Proc. IRE* , **50**, 2061–2070, (1962).
12. D. G. Aronson and H. F. Weinberger,*in Partial Differential Equations and Related Topics (J. A. Goldstein ed.), Lect. Notes Math.*, **446**, Springer–Verlag, 5–49, (1975).
13. D. G. Aronson and H. F. Weinberger ,*Adv. Math.*, **30**, 33–76, (1978).

14. S. Lie,*Uber integration durch bestimente integrale von einer klasse lineare partiellen differentialgleichungen*, **6**, Leipzig, 328–368, (1881).
15. G. W. Bluman and J. D. Cole, *J. Math. Mech.*, **18**, 1025–1042, (1969).
16. L. V. Ovsiannikov, *The Group Analysis of Differential Equations*, Nauka, Moscow (1978).
17. V. A. Dorodnitsyn, *USSR Comput. Math. and Math. Phys.*, **22**, 115–122, (1982).
18. M. I. Serov, *Ukrainian Math.*, **42**, 1370–1376, (1990).
19. D. J. Arrigo, P. Broadbridge and J.M. Hill, *IMA J. Appl. Math.*, **52**, 1–24, (1994).
20. P. A. Clarkson and E. L. Mansfield, *Physica D*, **70**, 250–288, (1993).
21. W. Fushchych, M. Serov and L. Tulupova, *Proc. Ukr. Acad. Sci.*, **4**, 23–27, (1993).
22. D. J. Arrigo and J. M. Hill, *Studies in Appl. Math.*, **94**, 21–39, (1995).
23. W. I. Fushchich and N. I. Serov, *Dokl. Akad. Nauk. Ukr. A*, **4**, 24–28, (1990).
24. I. Tsyfra and W. Fushchych, *Nonlinear Math. Phys.*, **2**, (1), 90–93, (1995).
25. M. J. Ablowitz and A. Zeppetella, *Bull. Math. Biol.*, **41**, 835–840, (1979).
26. B. Guo and Z.Chen, *J. Phys. A: Math. Gen.*, **24**, 645–650, (1991).
27. P. Kaliappan, *Physica D*, **11**, 368–374, (1984).
28. J. J. Herrera, A. Minzoni and R. Ondarza, *Physica D*, **57**, 249–266, (1992).
29. P. S. Bindu and M. Lakshmanan, *Proc. Inst. Math. NAS Ukr.*, **43** 1, 36–48, (2002).
30. V. M. Kenkre, *Physica A*, **342**, 242–248, (2004).
31. V. M. Kenre and M. N. Kuperman, *Phys. Rev. E*, **67**, 051921, (2003).
32. G. Abramson, A. R. Bishop and V. M. Kenkre, *Phys. Rev. E*, **64**, 066615, (2001).
33. P. K. Brazhnik and J. J. Tyson, *SIAM J. Appl. Math.*, **60**, (2), 371–391, (1999).
34. F. Cariello and M. Tabor, *Physica D*, **39**, 77–94, (1989).
35. F. Cariello and M. Tabor, *Physica D*, **53**, 59–70, (1991).
36. P. C. Fife and J. B. McLeod, *Arch. Rat. Mech. Anal.*, **65**, 335–361, (1977).
37. K. P.Hadeler and F. Rothe, *J. Math. Biol.*, **2**, 251–269, (1975).
38. E. J. M. Veling, *Proc. Roy. Soc. Edin.*, **90**, 41–61, (1981).
39. E. M. Vorob'ev, *Acta Appl. Math.*, **24**, 1–24, (1991).
40. T. Kawahara and M. Tanaka, *Phys. Lett. A*, **97**, 311-314, (1983).
41. M. C. Nucci and P. A. Clarkson, *Phys. Lett. A*, **164**, 49–56, (1992).
42. Z. X. Chen and B. Y. Guo, *Phys. Lett. A*, **48**, 107–115, (1992).
43. P. A. Clarkson and E. L. Mansfield, in *Applications of Analytic and Geometric Methods to Nonlinear Differential Equations*, P. A. Clarkson ed., Kluwer, Dordrecht, (1993).
44. N. H. Ibragimov and C. Sophocleous, *Proc. Inst. Math. NAS Ukr.*, **50**, (1), 142–148, (2004).
45. E. Pucci and G. Saccomandi, *Phisica D*, **139**, 28–47, (2000).
46. N. N. Ianenko and B. L. Rozdestvenski,*Systems of Quasilinear Equations and Their Applications to Gasdynamics*, Amer. Math. Soc., Providence, Rhode Island, **55**, (1983).

SOME APPLICATIONS OF LINEAR RESPONSE THEORY TO MEDIA WITH MECHANICAL RELAXATION PHENOMENA.

A. CIANCIO* and V. CIANCIO**

*Department of Mathematics, University of Messina,
c.da Parpardo - s.ta Sperone 31 , Messina 98166, Italy
E-mail:* aciancio@unime.it, ** ciancio@unime.it*

F. FARSACI

*IPCF – C.N.R., Messina, Italy.
E-mail: farsaci@me.cnr.it*

The general linear response theory it applied for study mechanical relaxations phenomena in continuous media. It is shown that phenomenological and state coefficients, in viscoanelastic media can be obtained as function of quantities which are experimentally measurable. The behaviour of the medium at low and high frequency is taking in account. Therefore, by compared the solutions of the differential equations that describe relaxation phenomena in the considered model to a relation that describe the same phenomena in linear response theory, it is possible to determine the aforementioned coefficients.

Keywords: Linear response theory, relaxation phenomena, phenomenological coefficients.

1. Linear response theory.

Let a generic system S be subject to a time dependent perturbation $f(t)$ (*input*) and let $g(t)$ the time dependent function which represent the response (*output*) of the system S to the perturbation $f(t)$ (see Fig.1).

Generally a physical problem is to investigate some characteristics of the system S when $f(t)$ and $g(t)$ are known functions. This is possible if a relation between $f(t)$ and $g(t)$ is obtained; this relation can be represented by an operator as

$$\Omega(t, a_1(t), a_2(t), \ldots) \tag{1}$$

function of the time and parameters $a_1(t), a_2(t), \ldots$ that describe the phenomena to be investigated.

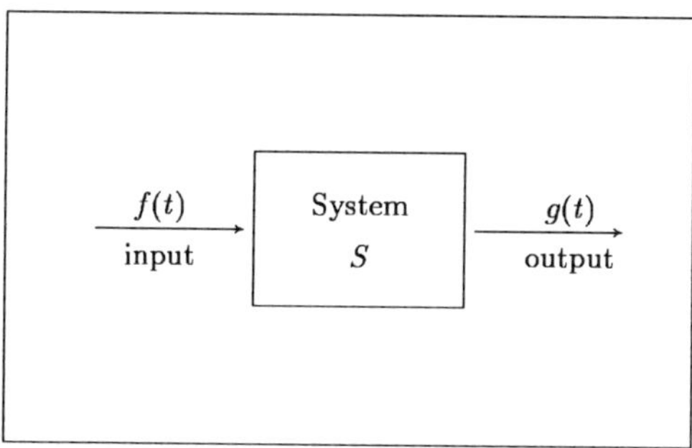

Fig. 1.   Schematic response experiment.

Mathematically this problem can be expressed by the relation

$$g(t) = \Omega(t, a_1(t), a_2(t), \ldots) \bullet f(t) \tag{2}$$

in which $\Omega(t, a_1(t), a_2(t), \ldots)$ is the unknown function to determine and the symbol "$\bullet$" means that when Ω acts on $f(t)$ the result will be $g(t)$.

Generally the determination of $\Omega(t, a_1(t), a_2(t), \ldots)$ is very difficult, nevertheless there exists a set of systems[11] that allow us to determine this operator : *the Linear Systems Shifth Invarant* (LSSI). A system S is linear shift invariant if between the input signal (f) and the output signal (g) one has:

$$\left. \begin{array}{c} f_1(t) \to g_1(t) \\ f_2(t) \to g_2(t) \end{array} \right| \Rightarrow f_1(t) + g_1(t) \to f_2(t) + g_2(t) , \tag{3}$$

$$f(t) \to g(t) \Rightarrow f(t - \tau) \to g(t - \tau) . \tag{4}$$

For a LSSI it can be shown that the following convolution relation between $f(t)$ and $g(t)$ is valid

$$g(t) = f(t) \otimes h(t) = \int_{-\infty}^{+\infty} f(t_1) h(t - t_1) \, dt_1 \tag{5}$$

¿From (5) and taking in account convolution theorem it follows:

$$\mathcal{F}(g) = \mathcal{F}(f)\mathcal{F}(h) , \tag{6}$$

where $\mathcal{F}(\cdot)$ is the Fourier transform.

We set

$$\mathcal{F}(g) = G\left(\omega\right) = \int\limits_{-\infty}^{+\infty} e^{-i\omega t} g\left(t\right) dt \,, \tag{7}$$

$$\mathcal{F}(f) = F\left(\omega\right) = \int\limits_{-\infty}^{+\infty} e^{-i\omega t} \left(t\right) dt \,, \tag{8}$$

$$\mathcal{F}(h) = H\left(\omega\right) = \int\limits_{-\infty}^{+\infty} e^{-i\omega t} h\left(t\right) dt \,, \tag{9}$$

where $H\left(\omega\right)$ is transfer function.

By using (7)-(9)from (6) we have:

$$H(\omega) = \frac{G(\omega)}{F(\omega)} \,, \tag{10}$$

and consequently:

$$h(t) = \mathcal{F}^{-1}(H) = \mathcal{F}^{-1}\left(\frac{G}{F}\right) \,, \tag{11}$$

where $\mathcal{F}^{-1}(\cdot)$ is the inverse Fourier transform.

We will show as $h(t)$ can be obtained by the experimental evaluations of $f(t)$ and $g(t)$.[11] In the following we assume that the temperature is constant.

¡it can be shown that in a LSSI, to an harmonic input corresponds an harmonic output of the same frequency but with different amplitude and phase which depend on angular frequency of input, so if we utilize an harmonic input for $f(t)$, i.e.

$$f(t) = A e^{i\omega_0 t} \tag{12}$$

the output $g(t)$ can be expressed as

$$g(t) = B\left(\omega_0\right) e^{i(\omega_0 t + \phi(\omega_0))} \,. \tag{13}$$

Assuming assume an extensive variable as input (cause) and evaluate the corresponding intensive variable as output (effect), we are able to study the aforementioned relaxation phenomena[4-.9]

By calculation of the Fourier Transform of input (12) and output (13):

$$\begin{cases} \mathcal{F}\left(\omega\right) = \mathcal{F}\left\{A e^{i\omega_0 t}\right\} = 2\pi A \delta\left(\omega - \omega_0\right) \,, \\[2em] G\left(\omega\right) = \mathcal{F}\left\{B\left(\omega_0\right) e^{i(\omega_0 t + \phi(\omega_0))}\right\} = 2\pi B\left(\omega_0\right) e^{i\phi(\omega_0)} \delta\left(\omega - \omega_0\right) \end{cases} \tag{14}$$

from (10) it is easy to obtain the transfer function :

$$H\left(\omega\right) = \frac{G\left(\omega\right)}{F\left(\omega\right)} = \frac{B\left(\omega_0\right)}{A}e^{i\phi\left(\omega_0\right)} = const. \tag{15}$$

This transfer function allows us to introduce a complex quantity

$$\Omega\left(\omega\right) = \Omega_1\left(\omega\right) + i\Omega_2\left(\omega\right) = H\left(\omega\right) \tag{16}$$

with real and imaginary part given by

$$\begin{cases} \Omega_1 = \dfrac{B\left(\omega_0\right)}{A}\cos\phi\left(\omega_0\right) = const.\,, \\[4mm] \Omega_2 = \dfrac{B\left(\omega_0\right)}{A}\sin\phi_0\left(\omega\right) = const.. \end{cases} \tag{17}$$

In a experimental physical contest the quantities Ω_1 and Ω_2 are called **storage modulus** and **loss modulus**,[10][12] and it is possible to show that they are related to not dissipative phenomena and to dissipative one, respectively. Their physical meaning depend on physical meaning of the amplitude A and $B(\omega)$. Moreover these two quantities are directly experimental measurable as function of angular frequency.[12]

In linear region, by experimental observations, we assume that exist two values ω_R and ω_U such that:

$$\varphi\left(\omega_R\right) = \varphi_R \cong 0\,, \quad \Omega_1\left(\omega\right) = \Omega_1\left(\omega_R\right) = \Omega_{1R} \cong const. \quad \text{for } \omega \leq \omega_R$$

$$\varphi\left(\omega_U\right) = \varphi_U \cong 0\,, \quad \Omega_1\left(\omega\right) = \Omega_1\left(\omega_U\right) = \Omega_{1U} \cong const. \quad \text{for } \omega \geq \omega_U$$

Consequently, from $(17)_2$ we have:

$$\Omega_2\left(\omega\right) = \Omega_2\left(\omega_R\right) = \Omega_{2R} \cong 0 \qquad for \quad \omega \leq \omega_R$$

$$\Omega_2\left(\omega\right) = \Omega_2\left(\omega_U\right) = G_{2U} \cong 0 \qquad for \quad \omega \geq \omega_U$$

We shall call low frequency linear region and high frequency linear region that for which it results $\omega\sigma \ll 1$ or $\omega\sigma \gg 1$, respectively, where σ is relaxation time of the medium.

The method of linear response theory was been applied[1] to electromagnetic media with dielectric relaxation phenomena.

In the following sections we will apply this approach to linear mechanical and to obtain phenomenological and state coefficients as function of the aforementioned storage and loss modulus and so experimentally valuables.

2. Phenomenological approach to mechanical relaxation phenomena.

Let a medium subject to a shear deformation (extensive variable = cause) of the form (12) (input) and assume that just one component of the stain is different from zero, which we indicate with $\varepsilon^*(t)$, i.e.[10]

$$\varepsilon^*(t) = \varepsilon_0 e^{i\omega t} \tag{18}$$

where $f = \varepsilon^*$ and $A = \varepsilon_0$.

Of course the medium will oppose by a stress (intensive variable = effect) of the form (13) (output),i.e.:

$$\tau^*(t) = \tau_0(\omega)\, e^{i(\omega t + \phi(\omega))} \tag{19}$$

Taking in account (17) we have

$$\begin{cases} \Omega_1(\omega) = G_1(\omega) = \dfrac{\tau_0(\omega)}{\varepsilon_0} \cos\phi(\omega) \\[4mm] \Omega_2(\omega) = G_2(\omega) = \dfrac{\tau_0(\omega)}{\varepsilon_0} \sin\varphi(\omega) \end{cases} . \tag{20}$$

The physical content of (18), and (19) can be expressed by their real or coefficient of imaginary part; since nothing changes by the choice of one of these, for our purpose it is useful to consider the coefficients of imaginary part which we indicate with $\varepsilon(t)$ and $\tau(t)$, respectively.

By virtue of (20), we have:

$$\begin{cases} \varepsilon(t) = \varepsilon_0 \sin\omega t \\[4mm] \tau(t) = [\varepsilon_0 G_1(\omega)] \sin\omega t + [\varepsilon_0 G_2(\omega)] \cos\omega t \end{cases} . \tag{21}$$

2.1. *Viscoanelastic media of order one with memory.*

It has been shown[6–8] that, introducing one thermodynamical tensorial internal variable in entropy function (medium of order one) the behavior of an isotropic viscoanelastic medium with memory can be decribed with the following rheological equation if cross-effect between viscous and inelastic flows are neglected:

$$\frac{d\tilde{\tau}_{\alpha\beta}}{dt} + R_0^{(\tau)}\tilde{\tau}_{\alpha\beta} = R_0^{(\varepsilon)}\tilde{\varepsilon}_{\alpha\beta} + R_1^{(\varepsilon)}\frac{d\tilde{\varepsilon}_{\alpha\beta}}{dt} + R_2^{(\varepsilon)}\frac{d^2\tilde{\varepsilon}_{\alpha\beta}}{dt^2} , \tag{22}$$

where $\tilde{\tau}_{\alpha\beta}$ and $\tilde{\varepsilon}_{\alpha\beta}$ are the deviators of the stress and strain tensors, respectively.

The following hold

$$\begin{cases} R_0^{(\tau)} = a^{(1,1)}\eta_s^{(1,1)} \\[2mm] R_0^{(\varepsilon)} = a^{(0,0)}\left(a^{(1,1)} - a^{(0,0)}\right)\eta_s^{(1,1)} \\[2mm] R_1^{(\varepsilon)} = a^{(0,0)} + a^{(1,1)}\eta_s^{(1,1)}\eta_s^{(0,0)} \\[2mm] R_2^{(\varepsilon)} = \eta_s^{(0,0)} \end{cases} \qquad (23)$$

in which $a^{(0,0)}$ and $a^{(1,1)}$ are state coefficients while $\eta_s^{(0,0)}$ and $\eta_s^{(1,1)}$ are phenomenological coefficients related to the following physical phenomena

$$\begin{array}{ccl} a^{(0,0)} & \Rightarrow & elasticity \\ a^{(1,1)} & \Rightarrow & inelasticity \\ \eta_s^{(0,0)} & \Rightarrow & viscosity \\ \eta_s^{(1,1)} & \Rightarrow & fluidity \end{array} \qquad (24)$$

Therefore, in[7,8] it was shown that, from the positive character of the entropy production, the following inequalities hold:

$$a^{(0,0)} > 0\,, \quad a^{(1,1)} > 0\,, \quad \eta_s^{(1,1)} > 0\,, \quad \eta_s^{(0,0)} > 0\,, \qquad (25)$$

If we consider the case for which just one component of the strain and stress are different from zero, for example $\tilde{\varepsilon}_{12} = \varepsilon, \tilde{\tau}_{12} = \tau$ the equation (22) becomes:

$$\frac{d\tau}{dt} + R_0^{(\tau)}\tau = R_0^{(\varepsilon)}\varepsilon + R_1^{(\varepsilon)}\frac{d\varepsilon}{dt} + R_2^{(\varepsilon)}\frac{d^2\varepsilon}{dt^2} \qquad (26)$$

Let the medium be subject to one-dimensional harmonic shear deformation of the form

$$\varepsilon = \varepsilon_0 \sin\omega t \qquad (27)$$

that substitute in (26) gives:

$$\frac{d\tau}{dt} + \frac{\tau}{\sigma} = \alpha\sin\omega t + \beta\cos\omega t\,, \qquad (28)$$

where

$$\alpha = \left(R_0^{(\varepsilon)} - \omega^2 R_2^{(\varepsilon)}\right)\varepsilon_0 \quad ; \quad \beta = R_1^{(\varepsilon)}\varepsilon_0\omega \quad ; \quad \sigma = \frac{1}{R_0^{(\tau)}}\,. \qquad (29)$$

In (28) σ is the relaxation time experimentally measurable.[12]

The solution of (29) is

$$\tau\left(t\right) = \frac{\alpha\sigma + \beta\omega\sigma^2}{1 + \omega^2\sigma^2}\sin\omega t + \frac{\beta\sigma - \alpha\omega\sigma^2}{1 + \omega^2\sigma^2}\cos\omega t \tag{30}$$

and comparing this equation with $(21)_2$ and taking in account (23) and (29), one has:

$$a^{(0,0)}\left(a^{(1,1)} - a^{(0,0)}\right)\eta_s^{(1,1)} = \frac{\eta_s^{(0,0)}\omega^2\sigma + G_1 - G_2\omega\sigma}{\sigma} \tag{31}$$

$$a^{(0,0)} + a^{(1,1)}\eta_s^{(1,1)}\eta_s^{(0,0)} = \frac{G_2 + G_1\omega\sigma}{\omega\sigma} \tag{32}$$

This is an algebraic system make by three equations with four unknown functions $a^{(0,0)}$, $a^{(1,1)}$, $\eta_s^{(0,0)}$, $\eta_s^{(1,1)}$ since G_1, G_2 and σ can be experimentally measured.

In[1] it was shown that the system complete with the following equation:

$$G_{1R/H} = \frac{a^{(0,0)}\left(a^{(1,1)} - a^{(0,0)}\right)}{a^{(1,1)}} \tag{33}$$

where we select the values G_{1R} or G_{1H} for the symbol $G_{1R/H}$ depending on we refer to low or high frequency, respectively.

$$a^{(0,0)}\left(\omega\right) = \frac{G_1\left(1 + \omega^2\sigma^2\right) - G_{1R/H}}{\omega^2\sigma^2}, \tag{34}$$

$$a^{(1,1)}\left(\omega\right) = \frac{1}{\omega^2\sigma^2}\left\{\frac{\left[G_1\left(1 + \omega^2\sigma^2\right) - G_{1R/H}\right]^2}{\left(G_1 - G_{1R/H}\right)\left(1 + \omega^2\sigma^2\right)}\right\}, \tag{35}$$

$$\eta_s^{(1,1)}\left(\omega\right) = \omega^2\sigma\left\{\frac{\left(G_1 - G_{1R/H}\right)\left(1 + \omega^2\sigma^2\right)}{\left[G_1\left(1 + \omega^2\sigma^2\right) - G_{1R/H}\right]^2}\right\}, \tag{36}$$

$$\eta_s^{(0,0)}\left(\omega\right) = \frac{G_{1R/H} + G_2\omega\sigma - G_1}{\omega^2\sigma}. \tag{37}$$

In this way we have obtained the variables $a^{(0,0)}$, $a^{(1,1)}$, $\eta_s^{(0,0)}$, $\eta_s^{(1,1)}$, for low and high frequency, as functions of the frequency dependent quantities G_1 and G_2 which are experimentally determinable while σ is calculated by specific considerations on G_1 and G_2.

3. Conclusion

From a theoretical viewpoint the above results give a concrete contribution to complete the system of indefinite equations of continuum mechanics and to the possibility of obtaining by integration (and under particular conditions) analytical solutions at various frequency of deformation.

On the other hand, the method introduced allows us to obtain results that can be used in particular continua to test the applicability of theoretical model considered in this paper (as it happens for other models for which the method allows a complete algebraic system to be integrated). It also allows to verify, by means of the same fundamental inequalities, the presence of no linear or dissipative effects (as it is shown in fig.2 for PolyIsobutilene). From a technological viewpoint phenomenological and state coefficients can be adopted as a new approach to characterize linear mechanical material behaviour.

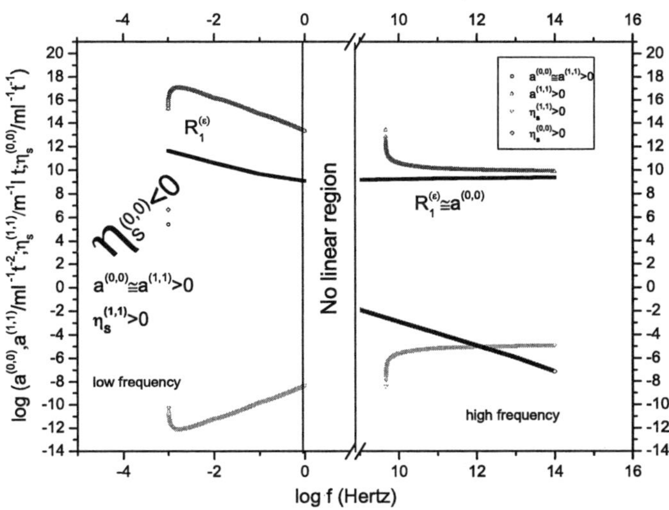

Fig. 2. Poly-isoButylene ($M.w. = 106 g/mol.$; $T_0 = 273 K$; $G_{1R} \approx 10^{5.4} Pa$; $G_{1U} \approx 10^{9.38} Pa$; $\sigma \approx 10^{-5} s.$).

References

1. V. Ciancio, F. Farsaci, G. Di Marco, *Physica B*, to appear.
2. V. Ciancio, A. V. Cimmelli, P. Vàn, *Mathematical and Computer Modelling*, **945**, 126 (2007).

3. V. Ciancio, M. Francaviglia, P. Rogolino, *Balkan Journal of Geometry and Its Applications* (BJGA), **9**,2, 1 (2004).
4. V. Ciancio, C. Cattani, *Conference on Applied Geometry-General Relativity and the Workshop of Global Analysis, Differenzial Geometry and Lie Algebras,* Grigorios Tsagas Ed., BSG, Proceedings 9, Bucharest , 249 (2003).
5. V. Ciancio, M. Francaviglia, *Balkan Journal of Geometry and Its Applications* (BJGA), **8**,1, 33 (2003).
6. V. Ciancio, J. Verhàs, *On the representation of dynamic degrees of freedom,* J. Non-Equilb. Thermodynamics, **18**, 39 (1993).
7. V. Ciancio, G. A. Kluitenberg, *Physica A* **99**, 592 (1979).
8. V. Ciancio V., G. A. Kuitenberg , *Physica A* **93**, 273 (1978).
9. G. D. C. Kuiken, *Thermodynamics of irreversible processes. Application to Diffusion and Rheology,* (John Wiley&Sons, Chichester - New York - Brisbane - Toronto - Singapore, 1994).
10. I. M. Ward, D. W. Hadley,*An Introduction to the mechanical properties of solid polymers,* (John Wiley-Sons, 1993).
11. D. C. Champeney,*Transform and their physical applications,*School of Mathematics and Physics Univesrity of East Anglia, Norwich, England (Academic Press, London - New York, 1973).
12. N. G. McCrum, B. E. Read, G.Williams,*Anelastic and Dielectric Effects in Polymeric Solids,* (John Wiley-Sons, London-New York, Sydney, 1967).

REDUCTION OF THE THREE-WAVE INTERACTION TO THE SIXTH PAINLEVÉ EQUATION

Robert Conte

Service de physique de l'état condensé (CNRS URA 2464),
CEA–Saclay, F–91191 Gif-sur-Yvette Cedex, France
E-mail: Robert.Conte@cea.fr

A. Michel Grundland

Centre de recherches mathématiques, Université de Montréal
Case postale 6128, Succursale Centre ville,
Montréal, Québec H3C 3J7, Canada
Département de mathématiques et d'informatique,
Université du Québec à Trois-Rivières
Case postale 500, Trois-Rivières, Québec G9A 5H7, Canada
E-mail: Grundlan@crm.umontreal.ca

Micheline Musette

Dienst Theoretische Natuurkunde, Vrije Universiteit Brussel,
Pleinlaan 2, B–1050 Brussels, Belgium
E-mail: MMusette@vub.ac.be

The sixth Painlevé equation P6 still lacks a second order matrix Lax pair which would be holomorphic in the four parameters of P6. We investigate here the construction of such a Lax pair from the reduction to P6 of some physical integrable system depending on space and time. Choosing this system as the three-wave resonant interaction, the reduction found by Kitaev yields a third order matrix Lax pair with explicit entries in terms of P6. In order to lower the matrix order to two, two methods can be used. The one due to Harnad reduces to the question of factorizing a rank two third order matrix into the product of two rectangular matrices.

Keywords: three-wave resonant interaction, reduction, sixth Painlevé equation.

1. Motivation

With a few notorious exceptions such as the Γ or Riemann ζ functions, all the so-called *special functions* are defined from an ordinary differential equation (ODE). Linear ODEs define among others the Gauss hypergeometric function and its degeneracies (Whittaker, Bessel, Hermite, Airy, etc). First order algebraic nonlinear ODEs only define the elliptic function. Second order algebraic nonlinear ODEs define one and only one master function, from the sixth Painlevé equation P6. It is therefore of the utmost importance to establish all the properties of the P6 function. Let us first recall the definition of P6, then state a missing property.

Consider a second order linear ODE for $\psi(t)$ with five Fuchsian singularities, one of them $t = u$ being apparent (i.e. the ratio of two linearly independent solutions remains single valued around it) and the four others having a crossratio x. The condition that the ratio ψ_1/ψ_2 of two linearly independent solutions be singlevalued when t goes around any of these singularities results in exactly one constraint between u and x, which is[1] that the apparent singularity u, considered as a function of the crossratio x, obeys the sixth Painlevé equation P6. In its normalized form (choice $(\infty, 0, 1, x)$ of the four nonapparent Fuchsian singularities), this ODE is[1]

$$\text{P6} \;:\; u'' = \frac{1}{2}\left[\frac{1}{u} + \frac{1}{u-1} + \frac{1}{u-x}\right]u'^2 - \left[\frac{1}{x} + \frac{1}{x-1} + \frac{1}{u-x}\right]u'$$
$$+ \frac{u(u-1)(u-x)}{x^2(x-1)^2}\left[\alpha + \beta\frac{x}{u^2} + \gamma\frac{x-1}{(u-1)^2} + \delta\frac{x(x-1)}{(u-x)^2}\right], \quad (1)$$

its four parameters $\alpha, \beta, \gamma, \delta$ representing the differences θ_j of the two Fuchs indices at the four nonapparent singularities $t = \infty, 0, 1, x$,

$$(2\alpha, -2\beta, 2\gamma, 1 - 2\delta) = (\theta_\infty^2, \theta_0^2, \theta_1^2, \theta_x^2). \quad (2)$$

If one also considers the x-dependence of the above mentioned wave function ψ, then ψ obeys a system of two linear equations[1–3]

$$\partial_t^2\psi + (S/2)\psi = 0, \quad \partial_x\psi + C\partial_t\psi - (1/2)C_t\psi = 0, \quad (3)$$

in which

$$-C = \frac{t(t-1)(u-x)}{(t-u)x(x-1)}, \quad \beta_1 = -\frac{x(x-1)}{2(u-x)}, \quad \beta_0 = -u + \frac{1}{2}, \quad (4)$$

$$-\frac{S}{2} = \frac{3/4}{(t-u)^2} + \frac{\beta_1 u' + \beta_0}{(t-u)t(t-1)} + \frac{[(\beta_1 u')^2 - \beta_0^2]\dfrac{u-x}{u(u-1)} + f_G(u)}{t(t-1)(t-x)} \quad (5)$$

$$+ f_G(t), \quad f_G(z) = \frac{a}{z^2} + \frac{b}{(z-1)^2} + \frac{c}{(z-x)^2} + \frac{d}{z(z-1)}, \quad (6)$$

$$(2\alpha, -2\beta, 2\gamma, 1 - 2\delta) = (4(a + b + c + d + 1), 4a + 1, 4b + 1, 4c + 1). \quad (7)$$

The commutativity condition of (3)

$$X \equiv S_x + C_{ttt} + CS_t + 2C_tS = 0 \quad (8)$$

is then equivalent to u being a solution of the P6 equation, i.e. (3) defines a scalar second order Lax pair for P6.

The unpleasant feature of the apparent singularity $t = u$ in (3) can be removed by considering a matrix Lax pair instead of a scalar one

$$\partial_x \Psi = L\Psi, \quad \partial_t \Psi = M\Psi, \quad L_t - M_x + LM - ML = 0. \tag{9}$$

There indeed exists a choice of L^4 for which the monodromy matrix M is defined as the sum of four Fuchsian singularities $t = \infty, 0, 1, x$,

$$M = \frac{M_0(x)}{t} + \frac{M_1(x)}{t-1} + \frac{M_x(x)}{t-x}, \quad M_\infty + M_0 + M_1 + M_x = 0. \tag{10}$$

However, this choice of L implies a meromorphic dependence of L on θ_∞ (details in Refs. 5,6), while u'' in the P6 equation is holomorphic in θ_∞.

The motivation of the present work is to build a second order matrix Lax pair (L, M) with exactly four Fuchsian singularities and a holomorphic dependence on the four θ_j's. In this short paper we summarize what has been achieved so far, the details being available in Refs. 6,7.

Our starting point will not be a mathematical system but a "simple" physical system governed by partial differential equations (PDE) enjoying two properties: (i) to admit a Lax pair, (ii) to admit a reduction to the generic P6 equation. By performing the reduction on the PDE Lax pair, there is a hope to obtain a Lax pair for P6 enjoying the desired property.

Several physical systems admit a reduction to P6, e.g.

(1) The three-wave resonant interaction system (3WRI)[8] in one space and one time dimensions, the three waves having a zero total impulsion $k_1 + k_2 + k_3 = 0$, a zero total pulsation $\omega_1 + \omega_2 + \omega_3 = 0$, and different group velocities c_j.
(2) The Maxwell-Bloch system,[9] whose Lax pair has second order.
(3) The Ernst equation,[10] itself a reduction of the Einstein equations.

2. The three-wave resonant interaction system

Let us consider the first of these systems. The three resonant waves in one space dimension x can be mathematically described[8] by six coupled partial differential equations in three complex amplitudes u_j,

$$\begin{cases} u_{j,t} + c_j u_{j,x} - i\overline{u}_k \overline{u}_l = 0, \\ \overline{u}_{j,t} + c_j \overline{u}_{j,x} + i u_k u_l = 0, \quad i^2 = -1, \end{cases} \tag{11}$$

in which (j, k, l) denotes any permutation of $(1, 2, 3)$, c_j are the constant values of the three different group velocities, $(c_2 - c_3)(c_3 - c_1)(c_1 - c_2) \neq 0$.

This system admits a third order Lax pair.[8] In the traceless zero curvature representation, this is given by[11]

$$\rho = -\frac{c_3 - c_1}{c_2 - c_3}, \quad \sigma = \frac{c_1 - c_2}{c_3 - c_1}, \tag{12}$$

$$L = \frac{i\lambda}{c_1 - c_2}\begin{pmatrix} -1+2\rho & 0 & 0 \\ 0 & 2-\rho & 0 \\ 0 & 0 & -1-\rho \end{pmatrix} + \frac{i}{c_1 - c_2}\begin{pmatrix} 0 & -\sigma\rho u_3 & \sigma\rho\overline{u}_2 \\ \sigma\overline{u}_3 & 0 & -\sigma u_1 \\ -u_2 & -\overline{u}_1 & 0 \end{pmatrix}, \tag{13}$$

$$M = \frac{i\lambda}{c_1 - c_2}\begin{pmatrix} c_1 - 2c_2\rho & 0 & 0 \\ 0 & -2c_1 + c_2\rho & 0 \\ 0 & 0 & c_1 + c_2\rho \end{pmatrix}$$

$$+ \frac{i}{c_1 - c_2}\begin{pmatrix} 0 & c_3\sigma\rho u_3 & -c_2\sigma\rho\overline{u}_2 \\ -c_3\sigma\overline{u}_3 & 0 & c_1\sigma u_1 \\ c_2 u_2 & c_1\overline{u}_1 & 0 \end{pmatrix}, \tag{14}$$

$$[\partial_x - L, \partial_t - M] = 0, \tag{15}$$

in which the spectral parameter λ is an arbitrary complex constant.

3. A reduction to P6

The reductions to six-dimensional ordinary differential systems have been investigated by many authors,[12-14] and, like for the Maxwell-Bloch system and the Ernst equation, their integration[6,12-14] involves most of the six Painlevé functions.

A thorough search by group theretical methods[14] has led to only one reduction to P6, which is,[12,13]

$$\begin{cases} \zeta = \dfrac{x}{t}, \quad \beta_1 + \beta_2 + \beta_3 = 0, \\ u_j(x,t) = (t(c_j - \zeta))^{-1+i\beta_j}\,\psi_j, \\ \overline{u}_j(x,t) = (t(c_j - \zeta))^{-1-i\beta_j}\,\overline{\psi}_j, \end{cases} \tag{16}$$

in which β_j are constants. The reduced system is

$$\begin{cases} \dfrac{d}{d\zeta}\psi_j = i(c_j - \zeta)^{-i\beta_j}(c_k - \zeta)^{-1-i\beta_k}(c_l - \zeta)^{-1-i\beta_l}\overline{\psi}_k\overline{\psi}_l, \\ \dfrac{d}{d\zeta}\overline{\psi}_j = -i(c_j - \zeta)^{i\beta_j}(c_k - \zeta)^{-1+i\beta_k}(c_l - \zeta)^{-1+i\beta_l}\psi_k\psi_l, \end{cases} \tag{17}$$

in which (j,k,l) denotes any permutation of $(1,2,3)$. As noticed by Kitaev, ζ, c_1, c_2, c_3 only contribute by their crossratio, so this six-dimensional system depends on two parameters β_j.

The reduced traceless Lax pair $(\mathcal{L}, \mathcal{M})$ in zero curvature representation

$$[\partial_\zeta - \mathcal{L}, \partial_\mu - \mathcal{M}] = 0, \tag{18}$$

depends on the constant spectral parameter μ,

$$\mathcal{L} = \frac{i}{c_1 - c_2}\mu \begin{pmatrix} -1+2\rho & 0 & 0 \\ 0 & 2-\rho & 0 \\ 0 & 0 & -1-\rho \\ +\frac{i}{c_1-c_2}\times \end{pmatrix}$$

$$\times \begin{pmatrix} 0 & -\sigma\rho\psi_3(c_3-\zeta)^{-1+\beta_3 i} & \sigma\rho\overline{\psi}_2(c_2-\zeta)^{-1-\beta_2 i} \\ \sigma\overline{\psi}_3(c_3-\zeta)^{-1-\beta_3 i} & 0 & -\sigma\psi_1(c_1-\zeta)^{-1+\beta_1 i} \\ -\psi_2(c_2-\zeta)^{-1+\beta_2 i} & -\overline{\psi}_1(c_1-\zeta)^{-1-\beta_1 i} & 0 \end{pmatrix},$$

$$\tag{19}$$

$$\mathcal{M} = \frac{i\rho(c_2-\zeta)}{c_1-c_2}\begin{pmatrix} -2 & 0 & 0 \\ 0 & 1 & 0 \\ 0 & 0 & 1 \end{pmatrix} + \frac{i(c_1-\zeta)}{c_1-c_2}\begin{pmatrix} 1 & 0 & 0 \\ 0 & -2 & 0 \\ 0 & 0 & 1 \end{pmatrix}$$

$$+\frac{i}{3}\mu^{-1}\begin{pmatrix} \beta_2-\beta_3 & 0 & 0 \\ 0 & \beta_3-\beta_1 & 0 \\ 0 & 0 & \beta_1-\beta_2 \end{pmatrix} \tag{20}$$

$$-\frac{i}{c_1-c_2}\mu^{-1}\begin{pmatrix} 0 & -\sigma\rho\psi_3(c_3-\zeta)^{\beta_3 i} & \sigma\rho\overline{\psi}_2(c_2-\zeta)^{-\beta_2 i} \\ \sigma\overline{\psi}_3(c_3-\zeta)^{-\beta_3 i} & 0 & -\sigma\psi_1(c_1-\zeta)^{\beta_1 i} \\ -\psi_2(c_2-\zeta)^{\beta_2 i} & -\overline{\psi}_1(c_1-\zeta)^{-\beta_1 i} & 0 \end{pmatrix}.$$

The singularities of the matrix $\mathcal{M}$ in the complex spectral parameter are $\mu = 0$ (of the Fuchsian type) and $\mu = \infty$ (of the nonFuchsian type).

The Fuchsian singularity in $\mathcal{M}$ allows one to generate the first integrals. Indeed, denoting $\mathcal{M}_{-1}$ the residue of the matrix $\mathcal{M}$ at the Fuchsian singularity $\mu = 0$,

$$\mathcal{M} = \mathcal{M}_{-1}\mu^{-1} + \mathcal{M}_0, \tag{21}$$

the invariants of the residue $\mathcal{M}_{-1}$ are constants of the motion,

$$\det(\mathcal{M}_{-1} - z) = -z^3 - \left(K_1 + \frac{\beta_1^2 + \beta_2^2 + \beta_3^2}{6}\right)z$$

$$+ 2i\left(K_2 - \frac{(\beta_2-\beta_3)(\beta_3-\beta_1)(\beta_1-\beta_2)}{54}\right), \tag{22}$$

in which K_1, K_2 denote the only two first integrals[14]

$$N = (c_2-c_3)(c_3-c_1)(c_1-c_2), \tag{23}$$

$$NK_1 = (c_2-c_3)\psi_1\overline{\psi}_1 + (c_3-c_1)\psi_2\overline{\psi}_2 + (c_1-c_2)\psi_3\overline{\psi}_3, \tag{24}$$

$$NK_2 = (1/2)\psi_1\psi_2\psi_3(c_1-\zeta)^{\beta_1 i}(c_2-\zeta)^{\beta_2 i}(c_3-\zeta)^{\beta_3 i}$$

$$+(1/2)\overline{\psi}_1\overline{\psi}_2\overline{\psi}_3(c_1-\zeta)^{-\beta_1 i}(c_2-\zeta)^{-\beta_2 i}(c_3-\zeta)^{-\beta_3 i}$$

$$+(1/6)\left((\beta_2-\beta_3)(c_2-c_3)\psi_1\overline{\psi}_1 + (\beta_3-\beta_1)(c_3-c_1)\psi_2\overline{\psi}_2\right.$$

$$\left. + (\beta_1-\beta_2)(c_1-c_2)\psi_3\overline{\psi}_3\right). \tag{25}$$

These first integrals only depend on the four quantities (ρ_j, χ) defined by

$$\psi_j = \rho_j e^{i\varphi_j}, \quad \overline{\psi}_j = \rho_j e^{-i\varphi_j}, \quad \chi = \sum_j (\varphi_j + \beta_j \log(c_j - \zeta)), \qquad (26)$$

the expressions for K_1, K_2 being

$$N K_1 = (c_2 - c_3)\rho_1^2 + (c_3 - c_1)\rho_2^2 + (c_1 - c_2)\rho_3^2, \qquad (27)$$

$$N K_2 = \rho_1 \rho_2 \rho_3 \cos\chi + \frac{1}{6}\sum_j (\beta_k - \beta_l)(c_k - c_l)\rho_j^2. \qquad (28)$$

Moreover, the four-dimensional system for (ρ_j, χ) is closed. This allows one to discard the three variables φ_j, remembering only their first derivatives

$$\varphi_j' = \frac{\rho_k \rho_l}{(c_k - \zeta)(c_l - \zeta)\rho_j} \cos\chi, \qquad (29)$$

and to focus on the closed fourth order system

$$\rho_j' = \frac{\rho_k \rho_l}{(c_k - \zeta)(c_l - \zeta)} \sin\chi, \qquad (30)$$

$$\chi' = \sum_j \left(\frac{\rho_k \rho_l}{(c_k - \zeta)(c_l - \zeta)\rho_j} \cos\chi - \frac{\beta_j}{c_j - \zeta} \right), \qquad (31)$$

which admits the two first integrals (27)–(28).

The counting now suits the integration with P6. Indeed, the P6 equation has second order and four parameters, and the system (30)–(31) has fourth order, two parameters and two algebraic first integrals.

The explicit integration has been performed in Refs. 6,14, and we now summarize it. The binomial second order second degree algebraic ODEs

$$Y''^2 = F(Y', Y, \zeta), \qquad (32)$$

have been "classified",[15] i.e. all such ODEs with the Painlevé property have been enumerated and integrated. One of the canonical equations so isolated is the one labeled SD.I.a [15, Eq. (5.4)],

$$-x^2(x-1)^2 y''^2 - 4y'(xy' - y)^2 + 4y'^2(xy' - y)$$

$$+A_0 y'^2 + A_2(xy' - y) + \left(A_3 + \frac{A_0^2}{4} \right) y' + A_4 = 0. \qquad (33)$$

The three variables ρ_j^2 are linear in y and y',

$$\rho_j^2 = \frac{(c_1 - c_2)(c_2 - c_3)(c_j - \zeta)}{c_2 - \zeta} y' - (c_3 - c_1)(c_j - c_2)y$$

$$- (c_j - c_k)(c_j - c_l)(1 - \delta_{j,2})\frac{K_1}{2} + (c_j - c_k)(c_j - c_l)\frac{\beta_2 \beta_{4-j}}{4}, \qquad (34)$$

with (j,k,l) a permutation of $(1,2,3)$, and the link between the constants

$$\begin{cases}
A_0 = -2K_1 - \dfrac{1}{2}\left(\beta_1^2 + \beta_2^2 + \beta_3^2\right), \\[2mm]
A_1 = \beta_1\left[\dfrac{\beta_2 - \beta_3}{3}K_1 + 4K_2 + \dfrac{1}{4}\beta_1^2(\beta_2 - \beta_3)\right] \\[2mm]
A_2 = \beta_2\left[\dfrac{\beta_3 - \beta_1}{3}K_1 + 4K_2 + \dfrac{1}{4}\beta_2^2(\beta_3 - \beta_1)\right], \\[2mm]
A_3 = \beta_3\left[\dfrac{\beta_1 - \beta_2}{3}K_1 + 4K_2 + \dfrac{1}{4}\beta_3^2(\beta_1 - \beta_2)\right], \\[2mm]
A_4 = -4K_2^2 - \dfrac{5\beta_2^2 + 2\beta_2\beta_3 + 2\beta_3^2}{18}K_1^2 - \dfrac{2(\beta_3 - \beta_1)}{3}K_1K_2 \\[2mm]
\quad - \beta_2^2\dfrac{5\beta_2^2 + 8\beta_2\beta_3 + 8\beta_3^2}{24}K_1 - \beta_2^2(\beta_3 - \beta_1)K_2 - \beta_2^4\dfrac{\beta_2^2 + 3\beta_2\beta_3 + 3\beta_3^2}{16}.
\end{cases} \tag{35}$$

The link of SD.I.a with P6 was mentioned by Chazy [16, Eq. B-V p. 340]. Its explicit integration with P6 was first performed in Ref. 17 and later simplified in Ref. 15. This link between P6 (1) and SD.I.a is [15, Eq. (5.19)],

$$\begin{cases}
y = \dfrac{x^2(x-1)^2}{4u(u-1)(u-x)}\left\{u' - \dfrac{u(u-1)}{x(x-1)}\right\}^2 + \dfrac{\Theta_\infty^2}{8}(1-2u) \\[2mm]
\quad + \dfrac{\theta_0^2}{8}\left(1 - 2\dfrac{x}{u}\right) + \dfrac{\theta_1^2}{8}\left(2\dfrac{x-1}{u-1} - 1\right) + \dfrac{\theta_x^2}{8}\left(1 - 2\dfrac{x(u-1)}{u-x}\right), \\[2mm]
\Theta_\infty = \theta_\infty + 1, \\[2mm]
2A_0 = \Theta_\infty^2 + \theta_0^2 + \theta_1^2 + \theta_x^2, \quad 4A_1 = -(\Theta_\infty^2 - \theta_0^2)(\theta_1^2 - \theta_x^2), \\[2mm]
4A_2 = -(\Theta_\infty^2 - \theta_x^2)(\theta_0^2 - \theta_1^2), \quad 4A_3 = (\Theta_\infty^2 - \theta_1^2)(\theta_0^2 - \theta_x^2), \\[2mm]
32A_4 = (\Theta_\infty^2 + \theta_x^2)(\theta_0^2 - \theta_1^2)^2 + (\Theta_\infty^2 - \theta_x^2)^2(\theta_0^2 + \theta_1^2).
\end{cases} \tag{36}$$

The elimination of (A_0, A_2, A_3, A_4) provides the link between the four essential parameters of the reduction (the two first integrals K_1, K_2 and the three constants β_j whose sum is zero), and the four monodromy exponents $(\theta_\infty, \theta_0, \theta_1, \theta_x)$ of P6,

$$\begin{cases}
4K_1 = -\left[\beta_1^2 + \beta_2^2 + \beta_3^2 + \Theta_\infty^2 + \theta_0^2 + \theta_1^2 + \theta_x^2\right], \\[2mm]
48K_2 = -\dfrac{(\Theta_\infty^2 - \theta_0^2)(\theta_1^2 - \theta_x^2)}{\beta_1} - \dfrac{(\Theta_\infty^2 - \theta_x^2)(\theta_0^2 - \theta_1^2)}{\beta_2} + \dfrac{(\Theta_\infty^2 - \theta_1^2)(\theta_0^2 - \theta_x^2)}{\beta_3} \\[2mm]
\quad + (\beta_1 - \beta_2)(\beta_2 - \beta_3)(\beta_3 - \beta_1), \\[2mm]
\beta_1\beta_2\beta_3(\beta_1^2 + \beta_2^2 + \beta_3^2) + 2\beta_1\beta_2\beta_3(\Theta_\infty^2 + \theta_0^2 + \theta_1^2 + \theta_x^2) \\[2mm]
\quad - 2\beta_1(\Theta_\infty^2\theta_0^2 + \theta_1^2\theta_x^2) - 2\beta_2(\Theta_\infty^2\theta_x^2 + \theta_0^2\theta_1^2) - 2\beta_3(\Theta_\infty^2\theta_1^2 + \theta_x^2\theta_0^2) = 0, \\[2mm]
-(\Theta_\infty^2 - \theta_x^2)^2(\theta_0^2 - \theta_1^2)^2 \\[2mm]
\quad - 2\beta_2^2\left[4\theta_1^4(\Theta_\infty^2 + \theta_x^2) + 4\theta_x^4(\theta_0^2 + \theta_1^2)\right. \\[2mm]
\quad\quad \left. + (\Theta_\infty^2 + \theta_x^2)(\theta_0^2 + \theta_1^2)(\Theta_\infty^2 + \theta_0^2 - 3\theta_1^2 - 3\theta_x^2)\right] \\[2mm]
\quad - \beta_2^4\left[(\Theta_\infty^2 + \theta_0^2 + \theta_1^2 + \theta_x^2)^2 - 2(\Theta_\infty^2 - \theta_x^2)(\theta_0^2 - \theta_1^2)\right] \\[2mm]
\quad + 4\beta_2^3\beta_3(\Theta_\infty^2 - \theta_x^2)(\theta_0^2 - \theta_1^2) \\[2mm]
\quad - 2\beta_2^4(\beta_2^2 + 2\beta_2\beta_3 + 2\beta_3^2)(\Theta_\infty^2 + \theta_0^2 + \theta_1^2 + \theta_x^2) - \beta_2^4(\beta_1^2 + \beta_3^2)^2 = 0.
\end{cases} \tag{37}$$

74

The above first three equations are invariant under both the ternary symmetry on β_j and the quaternary symmetry on $(\Theta_\infty, \theta_0, \theta_1, \theta_x)$.

From these results, one can establish the singlevaluedness [6, Section 6] (more precisely the absence of movable critical singularities) of all the elements in the reduced Lax pair (20).

4. Route from the three-wave system to a second order matrix Lax pair for P6

Knowing the third order matrix Lax pair (20), the question is now to build another matrix Lax pair which would have second order and, when expressed in terms of a solution u of P6, would hopefully be holomorphic in the four θ_j. Two observations are in order.

A first observation is that the only singularities in the complex μ plane admitted by the third order monodromy matrix $\mathcal{M}$ Eq. (20) are $\mu = 0$ (of the Fuchsian type) and $\mu = \infty$ (of the nonFuchsian type), i.e. exactly those of another third order matrix introduced[18] to describe the monodromy of a time-dependent Hamiltonian with three degrees of freedom, and later considered independently[19] from the point of view of its Laplace transform.

As a second observation, a duality has been established by two different methods (factorization of a residue,[18] Laplace transform in the μ space[19]) between the third order Lax pair of the above mentioned time-dependent Hamiltonian and a second order matrix Lax pair admitting as only singularities four Fuchsian points. This latter second order Lax pair, as expected, admits the generic P6 equation as its zero-curvature condition.

These two observations should have two consequences, yet unproven. (i) There should exist an identification between the two systems (reduced three-wave, time-dependent Hamiltonian). (ii) The third order matrix Lax pair Eq. (20) should have a dual, second order matrix Lax pair admitting P6 as its zero-curvature condition.

Let us first present the time-dependent Hamiltonian and its third order and second order matrix Lax pairs, then explain where the difficulty lies in the three-wave system.

5. A three degree of freedom Hamiltonian connected to P6

The time-dependent Hamiltonian with three degrees of freedom [18, Eq. (3.56)]

$$H(q_j, p_j, x) = \frac{1}{4}\left[g_{31}(x)a_{13}a_{31} + g_{23}(x)a_{23}a_{32} + g_{12}(x)a_{12}a_{21}\right] \quad (38)$$

with the notation

$$a_{12} = q_1 p_2 - q_2 p_1 + (\mu_1/q_1)q_2 + (\mu_2/q_2)q_1,$$
$$a_{21} = q_2 p_1 - q_1 p_2 + (\mu_2/q_2)q_1 + (\mu_1/q_1)q_2,$$
$$a_{13} = q_1 p_3 + q_3 p_1 - (\mu_1/q_1)q_3 + (\mu_3/q_3)q_1,$$
$$a_{31} = q_3 p_1 + q_1 p_3 - (\mu_3/q_3)q_1 + (\mu_1/q_1)q_3,$$
$$a_{23} = q_2 p_3 + q_3 p_2 - (\mu_2/q_2)q_3 + (\mu_3/q_3)q_2,$$
$$a_{32} = q_3 p_2 + q_2 p_3 - (\mu_3/q_3)q_2 + (\mu_2/q_2)q_3,$$
$$g_1 = 0, \ g_2 = 1, \ g_3 = x, \ g_{23} = \frac{1}{x-1}, \ g_{31} = \frac{1}{x}, \ g_{12} = 0, \qquad (39)$$

defines a six-dimensional first order system made of the six Hamilton equations in the canonical variables (q_j, p_j). This system admits the time-independent first integral

$$I = a_{13}a_{31} + a_{23}a_{32} + a_{12}a_{21} + 2(\mu_1^2 + \mu_2^2 + \mu_3^2), \qquad (40)$$

and the third order matrix Lax pair [18, Eqs. (3.62), (3.65)] (see also Ref. 19),

$$[\partial_x - \mathcal{L}_3, \partial_\lambda - \mathcal{M}_3] = 0,$$

$$\mathcal{L}_3 = -\lambda \begin{pmatrix} g_1' & 0 & 0 \\ 0 & g_2' & 0 \\ 0 & 0 & g_3' \end{pmatrix} + \frac{1}{2} \begin{pmatrix} 0 & g_{12}a_{12} & g_{31}a_{13} \\ g_{12}a_{21} & 0 & g_{23}a_{23} \\ g_{31}a_{31} & g_{23}a_{32} & 0 \end{pmatrix}$$

$$+ \frac{1}{2} \operatorname{diag} \left(g_{31} \left(\mu_1(q_3/q_1)^2 - \mu_3\right) + g_{12} \left(\mu_1(q_2/q_1)^2 - \mu_2\right), \right.$$
$$g_{23} \left(\mu_2(q_3/q_2)^2 - \mu_3\right) + g_{12} \left(\mu_2(q_1/q_2)^2 - \mu_1\right),$$
$$\left. g_{31} \left(\mu_3(q_1/q_3)^2 - \mu_1\right) + g_{23} \left(\mu_3(q_2/q_3)^2 - \mu_2\right)\right),$$

$$\mathcal{M}_3 = - \begin{pmatrix} g_1 & 0 & 0 \\ 0 & g_2 & 0 \\ 0 & 0 & g_3 \end{pmatrix} + \frac{\mathcal{R}_{-1}}{\lambda}, \ \mathcal{R}_{-1} = \frac{1}{2} \begin{pmatrix} 2\mu_1 & a_{12} & a_{13} \\ a_{21} & 2\mu_2 & a_{23} \\ a_{31} & a_{32} & 2\mu_3 \end{pmatrix}. \qquad (41)$$

The key property of this Hamiltonian is that the residue $\mathcal{R}_{-1}$, which has rank two, factorizes as[18]

$$\mathcal{R}_{-1} = FG, \ F = \frac{1}{\sqrt{2}} \begin{pmatrix} q_1 & p_1 - \mu_1/q_1 \\ q_2 & p_2 - \mu_2/q_2 \\ q_3 & -p_3 + \mu_3/q_3 \end{pmatrix}, \qquad (42)$$

$$G = \frac{1}{\sqrt{2}} \begin{pmatrix} p_1 + \mu_1/q_1 & p_2 + \mu_2/q_2 & p_3 + \mu_3/q_3 \\ -q_1 & -q_2 & -q_3 \end{pmatrix}. \qquad (43)$$

This factorization implies that this third order matrix Lax pair $(\mathcal{L}_3, \mathcal{M}_3)$ admits a dual, second order matrix Lax pair $(\mathcal{L}_2, \mathcal{M}_2)$ defined as [18, Eq. (3.55), (3.61)]

$$[\partial_x - \mathcal{L}_2, \partial_\Lambda - \mathcal{M}_2] = 0, \tag{44}$$

$$\mathcal{L}_2 = -\frac{R_x}{\Lambda - x}, \quad \mathcal{M}_2 = \frac{R_0}{\Lambda} + \frac{R_1}{\Lambda - 1} + \frac{R_x}{\Lambda - x}, \tag{45}$$

$$R_0 = -G \operatorname{diag}(1,0,0)F, \quad R_1 = -G \operatorname{diag}(0,1,0)F, \quad R_x = -G \operatorname{diag}(0,0,1)F,$$

$$R_\infty = -R_0 - R_1 - R_x = GF, \tag{46}$$

with the four residues explicitly given by

$$2R_\infty = \mu_1 + \mu_2 + \mu_3$$
$$+ \begin{pmatrix} q_1p_1 + q_2p_2 + q_3p_3 & p_1^2 + p_2^2 - p_3^2 - (\mu_1/q_1)^2 - (\mu_2/q_2)^2 + (\mu_3/q_3)^2 \\ -q_1^2 - q_2^2 + q_3^2 & -q_1p_1 - q_2p_2 - q_3p_3 \end{pmatrix},$$

$$2R_0 = \begin{pmatrix} -q_1p_1 & (\mu_1/q_1)^2 - p_1^2 \\ q_1^2 & q_1p_1 \end{pmatrix} - \mu_1, \tag{47}$$

$$2R_1 = \begin{pmatrix} -q_2p_2 & (\mu_2/q_2)^2 - p_2^2 \\ q_2^2 & q_2p_2 \end{pmatrix} - \mu_2, \tag{48}$$

$$2R_x = \begin{pmatrix} -q_3p_3 & -(\mu_3/q_3)^2 + p_3^2 \\ -q_3^2 & q_3p_3 \end{pmatrix} - \mu_3. \tag{49}$$

The zero-curvature conditions of $(\mathcal{L}_2, \mathcal{M}_2)$ and $(\mathcal{L}_3, \mathcal{M}_3)$ are both equivalent to the Hamilton equations derived from (38). Therefore, since the singularities of the monodromy matrix $\mathcal{M}_2$ in the complex plane of Λ are four Fuchsian singularities (located at $\Lambda = \infty, 0, 1, x$), the Hamilton equations of (38) can be explicitly integrated in terms of P6.[18,19] In particular, the invariants of the four residues are constants of the motion,

$$\operatorname{tr} R_\infty = \mu_1 + \mu_2 + \mu_3, \ \operatorname{tr} R_0 = -\mu_1, \ \operatorname{tr} R_1 = -\mu_2, \ \operatorname{tr} R_x = -\mu_3, \tag{50}$$

$$\det R_\infty = -\frac{I}{4} + \frac{1}{2}(\mu_1 + \mu_2 + \mu_3)^2, \det R_0 = \det R_1 = \det R_x = 0. \tag{51}$$

In order to find the general solution of the Hamilton equations, there is no need to perform any integration, it is sufficient to identify the coefficients of the matrix Lax pair (45) with those of an existing matrix Lax pair for the P6 equation (1). If one chooses for this Lax pair the one in Ref. 4, the result can be found in Ref. 18 but it has a meromorphic dependence on θ_∞. To avoid this meromorphic dependence, it would be better to convert the matrix pair (45) to scalar form, then to identify the result with the holomorphic Lax pair of Fuchs,[1] see indications in Ref. 7.

6. Transposition to the three-wave system

One has to establish the correspondance between the two sets of six variables (q_j, p_j) and $(\psi_j, \overline{\psi}_j)$. The identification of the invariants of the two residues of the third order matrix Lax pairs first yields (one should take care that the residue $\mathcal{R}_{-1}$ is not traceless))

$$\forall z : \; \det(\mathcal{M}_{-1} - z) = \det\left(\mathcal{R}_{-1} - \frac{\mu_1 + \mu_2 + \mu_3}{3} - z\right), \qquad (52)$$

i.e.

$$\begin{cases} K_1 + \dfrac{\beta_1^2 + \beta_2^2 + \beta_3^2}{6} = -\dfrac{I}{4} + \dfrac{(\mu_1 + \mu_2 + \mu_3)^2}{6}, \\ K_2 - \dfrac{(\beta_2 - \beta_3)(\beta_3 - \beta_1)(\beta_1 - \beta_2)}{54} \\ \qquad = -i\dfrac{(\mu_1 + \mu_2 + \mu_3)I}{24} + \dfrac{5i}{108}(\mu_1 + \mu_2 + \mu_3)^3. \end{cases} \qquad (53)$$

The difference in the nature of the involved constants makes the identification uneasy. Indeed, the Hamiltonian system has three fixed constants (μ_1, μ_2, μ_3) and one movable constant (the first integral I), while the reduced three-wave system has two fixed constants (two elements among the three β_j) and two movable constants (the two first integrals K_1, K_2).

After expressing the residue $\mathcal{M}_{-1}$ in terms of the P6 function u, one must also factorize this residue into a product similar to (43). Since this residue has rank three, one first lowers its rank to two by applying the transition matrix $P = \mu^a \mathcal{I}$ to the Lax pair, in which $\mathcal{I}$ is the identity matrix and a is any of the constant roots of the characteristic polynomial (22). The factorization of the resulting rank two matrix as

$$R = \mathcal{M}_{-1} - a = FG, \quad \operatorname{tr} \mathcal{M}_{-1} = 0, \qquad (54)$$

with F a $(3, 2)$ matrix and G a $(2, 3)$ matrix, both of rank two, is possible [20, §3.5.4] but it is not unique. In particular, if the elements of F and G are restricted to rational functions of the R'_{ij}s, the resulting elements of F and G depend on four arbitrary functions of the R'_{ij}s, with no specific direct criterium to choose them, and this difficulty is not yet overcome. However, with the definition (45)–(46), the invariants of the four residues are independent of the choice of the four arbitrary functions,

$$\operatorname{tr} R_\infty = 3a, \; \operatorname{tr} R_0 = R_{11}, \; \operatorname{tr} R_1 = R_{22}, \; \operatorname{tr} R_x = R_{33}, \qquad (55)$$

$$\det R_\infty = 3a^2 + Q_2, \; \det R_0 = \det R_1 = \det R_x = 0, \qquad (56)$$

$$a^3 + Q_2 a + Q_3 = 0. \qquad (57)$$

7. Conclusion

It should be worthwhile to consider another physical system, for instance one mentioned in section 1, which already admits a second order matrix Lax pair. This would bypass the difficulty described in previous section.

Another way out of the difficulty could be to perform a Laplace transform following the method in Ref. 19.

Acknowledgments

RC warmly thanks the organizers for the quite friendly atmosphere of the conference. This work was partially supported by the NSERC research grant of Canada (for AMG), the Tournesol grant no. T2003.09 between Belgium and France (for RC and MM), and CEA (for AMG and MM).

References

1. R. Fuchs, *C. R. Acad. Sc. Paris* **141**, 555–558 (1905).
2. R. Fuchs, *Math. Annalen* **63**, 301–321 (1907).
3. R. Garnier, *Ann. Éc. Norm.* **29**, 1–126 (1912).
4. M. Jimbo and T. Miwa, *Physica D* **2**, 407–448 (1981).
5. Lin R.-l., R. Conte and M. Musette, *J. Nonl. Math. Phys.* **10**, Supp. 2, 107–118 (2003).
6. R. Conte, A. M. Grundland and M. Musette, *J. Phys. A: Math. Gen.* **39**, 12115-12127 (2006).
7. R. Conte, preprint S2006/074, 6 pages (2006).
8. V.E. Zakharov and S. V. Manakov, *Pis'ma Zh. Eksp. Teor. Fiz.* **18**, 413–417 (1973).
9. H. Steudel and R. Meinel, *Physica D* **21**, 155–162 (1986).
10. G. Calvert and N. M. J. Woodhouse, *Class. Quant. Grav.* **13**, L33–L39 (1996).
11. M. J. Ablowitz and R. Haberman, *Phys. Rev. Lett.* **35**, 1185–1188 (1975).
12. A. Fokas, R. A. Leo, L. Martina and G. Soliani, *Phys. Lett. A* **115**, 329–332 (1986).
13. A. V. Kitaev, *J. Phys. A* **23**, 3453–3553 (1990).
14. L. Martina and P. Winternitz, *Annals of Physics* **196**, 231–277 (1989).
15. C. M. Cosgrove and G. Scoufis, *Stud. Appl. Math.* **88**, 25–87 (1993).
16. J. Chazy, *Acta Math.* **34**, 317–385 (1911).
17. F. J. Bureau, A. Garcet et J. Goffar, *Annali di Mat. pura ed appl.* **92**, 177–191 (1972).
18. J. Harnad, *Commun. Math. Phys.* **166**, 337–365 (1994).
19. M. Mazzocco, in *The Kowalevski property* CRM Proc. Lecture Notes **32** (Amer. Math. Soc., Providence, RI, 2002), pp. 219–238.
20. M. L. Mehta, *Matrix theory* (Les éditions de physique, Les Ulis, 1989).

THE D'ALEMBERT-LAGRANGE PRINCIPLE FOR GRADIENT THEORIES AND BOUNDARY CONDITIONS

H. GOUIN

Université d'Aix-Marseille, 13397 Marseille Cedex 20, France
E-mail: henri.gouin@univ-cezanne.fr

Dedicated to Prof. Antonio M. Greco

Motions of continuous media presenting singularities are associated with phenomena involving shocks, interfaces or material surfaces. The equations representing evolutions of these media are irregular through geometrical manifolds. A unique continuous medium is conceptually simpler than several media with surfaces of singularity. To avoid the surfaces of discontinuity in the theory, we transform the model by considering a continuous medium taking into account more complete internal energies expressed in gradient developments associated with the variables of state. Nevertheless, resulting equations of motion are of an higher order than those of the classical models: they lead to non-linear models associated with more complex integration processes on the mathematical level as well as on the numerical point of view. In fact, such models allow a precise study of singular zones when they have a non negligible physical thickness. This is typically the case for capillarity phenomena in fluids or mixtures of fluids in which interfacial zones are transition layers between phases or layers between fluids and solid walls. Within the framework of mechanics for continuous media, we propose to deal with the functional point of view considering globally the equations of the media as well as the boundary conditions associated with these equations. For this aim, we revisit the *d'Alembert-Lagrange principle of virtual works* which is able to consider the expressions of the works of forces applied to a continuous medium as a linear functional value on a space of test functions in the form of *virtual displacements*. At the end, we analyze examples corresponding to capillary fluids. This analysis brings us to numerical or asymptotic methods avoiding the difficulties due to singularities in simpler -but with singularities- models.

1. Introduction

A mechanical problem is generally studied through force interactions between masses located in material points: this Newton point of view leads together to the statistical mechanics but also to the continuum mechanics. The statistical mechanics is mostly precise but is in fact too detailed and in many cases huge calculations crop up. The continuum mechanics is an asymptotic notion coming from short range interactions between molecules. It follows a loose of information but a more efficient and directly computable theory. In the simplest case of continuum mechanics, residual information comes through stress tensor like Cauchy tensor[1,2]. The concept of stress tensor is so frequently used that it has become as natural as the notion of force. Nevertheless, tensor of contact couples can be investigated as in Cosserat medium[3] or configuration forces like in Gurtin approach[4] with edge interactions of Noll and Virga[5]. Stress tensors and contact forces are interrelated notions[6].

A fundamental point of view in continuum mechanics is: the Newton system for forces is equivalent to *the work of forces is the value of a linear functional of displacements*. Such a method due to Lagrange is dual of the system of forces due to Newton[7,8] and is not issued from a variational approach; the minimization of the energy coincides with the functional approach in a special variational principle only for some equilibrium cases.

The linear functional expressing the work of forces is related to the theory of distributions; a decomposition theorem associated with displacements (as test functions whose supports are C^∞ compact manifolds) uniquely determines a canonical zero order form *(separated form)* with respect both to the test functions and the transverse derivatives of contact test functions[9]. As Newton's principle is useless when we do not have any constitutive equation for the expression of forces, the linear functional method is useless when we do not have any constitutive assumption for the virtual work functional. The choice of the simple material theory associated with the Cauchy stress tensor corresponds with a constitutive assumption on its virtual work functional. It is important to notice that constitutive equations for the free energy χ and constitutive assumption for the virtual work functional may be incompatible[10] : for any *virtual* displacement ζ of an isothermal medium, the variation $-\delta\chi$ must be equal to the *virtual* work of internal forces $\delta\tau_{int}$. The equilibrium state is then obtained by the existence of a solution minimizing the free energy.

The equation of motion of a continuous medium is deduced from the *d'Alembert-Lagrange principle of virtual works* which is an extension of the principle in mechanics of systems with a finite number of degrees of freedom: *The motion is such that for any virtual displacement the virtual work of forces is equal to the virtual work of mass accelerations.*

Let us note: if the virtual work of forces is expressed in classical notations in the form

$$\delta\tau = \int\!\!\int\!\!\int_D \{\mathbf{f}.\boldsymbol{\zeta} + tr\left[(-p\,\mathbf{1} + 2\mu\,\nabla\mathbf{V}).\nabla\boldsymbol{\zeta}\right]\}\,dv + \int\!\!\int_S \mathbf{T}.\boldsymbol{\zeta}\,ds \quad (1)$$

from the d'Alembert-Lagrange principle, we obtain not only the equations of balance momentum for a viscous fluid in the domain D but also the boundary conditions on the border S of D. We notice that expression (1) is not the Frechet derivative of any functional expression.

If the free energy depends on the strain tensor F, then $\delta\tau$ must depend on $\nabla\boldsymbol{\zeta}$ and leads to the existence of the Cauchy stress tensor. If the free energy depends on the strain tensor F and on the overstrain tensor ∇F, then $\delta\tau$ must depend on $\nabla\boldsymbol{\zeta}$ and $\nabla^2\boldsymbol{\zeta}$.

Conjugated (or transposed) mappings being denoted by asterisk, for any vectors $\mathbf{a}, \mathbf{b}$, we write $\mathbf{a}^*\mathbf{b}$ for their *scalar product* (the line vector is multiplied by the column vector) and $\mathbf{ab}^*$ or $\mathbf{a}\otimes\mathbf{b}$ for their *tensor product* (the column vector is multiplied by the line vector). The product of a mapping $\mathbf{A}$ by a vector $\mathbf{a}$ is denoted by $\mathbf{A}\,\mathbf{a}$. Notation $\mathbf{b}^*\mathbf{A}$ means the covector $\mathbf{c}^*$ defined by the rule $\mathbf{c}^* = (\mathbf{A}^*\mathbf{b})^*$. The divergence of a linear transformation $\mathbf{A}$ is the covector $\mathrm{div}\,\mathbf{A}$ such that, for any constant vector $\mathbf{a}$, $(\mathrm{div}\,\mathbf{A})\,\mathbf{a} = \mathrm{div}\,(\mathbf{A}\,\mathbf{a})$. We introduce a Galilean or fixed system of coordinates (x^1, x^2, x^3) which is also denoted by $\mathbf{x}$ as Euler or spatial variables. If f is a real function of $\mathbf{x}$, $\dfrac{\partial f}{\partial\mathbf{x}}$ is the linear form associated with the gradient of f and $\dfrac{\partial f}{\partial x^i} = (\dfrac{\partial f}{\partial\mathbf{x}})_i$; consequently, $(\dfrac{\partial f}{\partial\mathbf{x}})^* = \mathrm{grad}\,f$. The identity tensor is denoted by $\mathbf{1}$.

Now, we present the method and its consequences in different cases of gradient theory. As examples, we revisit the case of Laplace theory of capillarity and the case of van der Waals fluids.

2. Virtual work of continuous medium

The motion of a continuous medium is classically represented by a continuous transformation φ of a three-dimensional space into the physical space. In order to describe the transformation analytically, the variables $\mathbf{X} = (X^1, X^2, X^3)$ which single out individual particles correspond to material or Lagrange variables. Then, the transformation representing the motion of a continuous medium is

$$\mathbf{x} = \varphi\,(\mathbf{X},t) \quad \text{or} \quad x^i = \varphi^i(X^1, X^2, X^3, t)\,, \quad i = 1, 2, 3$$

where t denotes the time. At t fixed the transformation possesses an inverse and continuous derivatives up to the second order except at singular sur-

82

faces, curves or points. Then, the diffeomorphism φ from the set D_0 of the particles into the physical space D is an element of a functional space $\wp$ of the positions of the continuous medium considered as a manifold with an infinite number of dimensions.

To formulate the d'Alembert-Lagrange principle of virtual works, we introduce the notion of *virtual displacements*. This is obtained by letting the displacements arise from variations in the paths of the particles. Let a one-parameter family of varied paths or *virtual motions* denoted by $\{\varphi_\eta\}$ and possessing continuous derivatives up to the second order and expressed analytically by the transformation

$$\mathbf{x} = \mathbf{\Phi}\,(\mathbf{X},t;\eta)$$

with $\eta \in O$, where O is an open real set containing 0 and such that $\mathbf{\Phi}\,(\mathbf{X},t;0) = \varphi\,(\mathbf{X},t)$ or $\varphi_0 = \varphi$ (the real motion of the continuous medium is obtained when $\eta = 0$). The derivation with respect to η when $\eta = 0$ is denoted by δ. Derivation δ is named *variation* and the *virtual displacement* is the variation of the position of the medium[11] . The virtual displacement is a tangent vector to $\wp$ in φ $(\delta\varphi \in T_\varphi(\wp))$. In the physical space, the *virtual displacement* $\delta\varphi$ is determined by the variation of each particle: the *virtual displacement of the particle* $\mathbf{x}$ is such that $\boldsymbol{\zeta} = \delta\mathbf{x}$ when $\delta\mathbf{X} = 0$, $\delta\eta = 1$ at $\eta = 0$; we associate the field of tangent vectors to D

$$\mathbf{x} \in D \to \boldsymbol{\zeta} = \psi(\mathbf{x}) \equiv \frac{\partial\mathbf{\Phi}}{\partial\eta}\,|_{\eta=0} \in T_\mathbf{x}(D)$$

where $T_\mathbf{x}(D)$ is the tangent vector bundle to D at $\mathbf{x}$. The concept of virtual

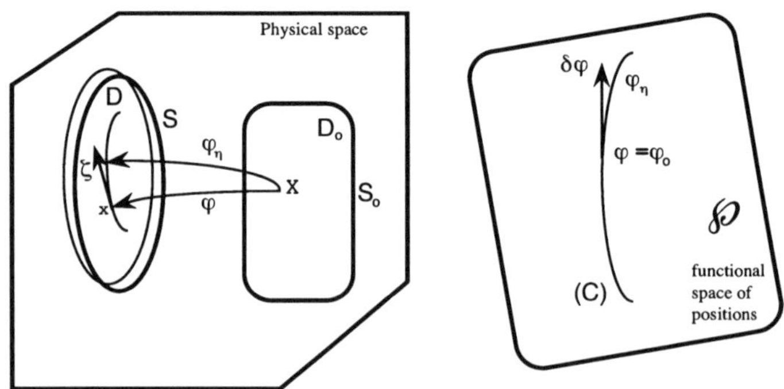

Fig. 1. The boundary S of D is represented by a thick curve and its variation by a thin curve. Variation $\delta\varphi$ of family $\{\varphi_\eta\}$ of varied paths belongs to $T_\varphi(\wp)$, tangent space to $(\wp)$ at φ.

work is purposed in the form:

The virtual work is a linear functional value of the virtual displacement,

$$\delta\tau = <\Im, \delta\varphi> \tag{2}$$

where $< .,. >$ denotes the inner product of $\Im$ and $\delta\varphi$; then, $\Im$ belongs to the cotangent space of $\wp$ at φ ($\Im \in T^*_\varphi(\wp)$).

In Relation (2), the medium in position φ is submitted to the covector $\Im$ denoting all the stresses; in the case of motion, we must add the inertial forces associated with the acceleration quantities to the volume forces. The d'Alembert-Lagrange principle of virtual works is expressed as:

For all virtual displacements, the virtual work is null.

Consequently, representation (2) leads to:

$$\forall\, \delta\varphi \in T_\varphi(\wp),\; \delta\tau = 0$$

Theorem: *If expression (2) is a distribution in a separated form, the d'Alembert-Lagrange principle yields the equations of motions and boundary conditions in the form* $\Im = 0$.

3. Some examples of linear functional of forces

Among all possible choices of linear functional of virtual displacements, we classify the following ones:

3.1. *Model of zero gradient*

3.1.1. *Model A.0*

The medium fills an open set D of the physical space and the linear functional is in the form

$$\delta\tau = \int\int\int_D F_i\, \zeta^i dv$$

where F_i ($i = 1, 2, 3$) denote the covariant components of the volume force **F** (including the inertial force terms) presented as a covector. The equation of the motion is

$$\forall\, \mathbf{x} \in D,\; F_i = 0 \Leftrightarrow \mathbf{F} = 0 \tag{3}$$

3.1.2. *Model B.0*

The medium fills a set D and the surface S is the boundary of D belonging to the medium; with the same notations as in section *3.1.1*, the linear

84

functional is in the form

$$\delta\tau = \int\int\int_D F_i\,\zeta^i dv + \int\int_S T_i\,\zeta^i ds \tag{4}$$

T_i are the components of the surface forces (tension) $\mathbf{T}$. From Eq. (4), we obtain the equation of motion as in Eq. (3) and the boundary condition,

$$\forall\,\mathbf{x}\in S,\ \ T_i = 0\ \Leftrightarrow\ \mathbf{T} = 0$$

,

3.2. *Model of first gradient*

3.2.1. *Model A.1*

With the previous notations, the linear functional is in the form

$$\delta\tau = \int\int\int_D \left(F_i\,\zeta^i - \sigma_i^j\,\zeta^i_{,j}\right) dv$$

where σ_i^j $(i,j = 1,2,3)$ are the components of the stress tensor σ. Stokes formula gets back to the model *B.0* in the separated form

$$\delta\tau = \int\int\int_D \left(F_i + \sigma_{i,j}^j\right)\zeta^i dv - \int\int_S n_j\sigma_i^j\,\zeta^i ds$$

where n_j $(j = 1,2,3)$ are the components of a covector which is the annulator of the vectors belonging to the tangent plane at the boundary S. It is not necessary to have a metric in the physical space; nevertheless, for the sake of simplicity it is convenient to use the Euclidian metric; the vector $\mathbf{n}$ of components n^j $(j = 1,2,3)$ represents the external normal to S relatively to D; the covector $\mathbf{n}^\star$ is associated with the components n_j. We deduce the equation of motion

$$\forall\,\mathbf{x}\in D,\ \ F_i + \sigma_{i,j}^j = 0\ \Leftrightarrow\ \mathbf{F} + \operatorname{div}\sigma = 0 \tag{5}$$

and the boundary condition

$$\forall\,\mathbf{x}\in S,\ \ n_j\,\sigma_i^j = 0\ \Leftrightarrow\ \mathbf{n}^\star\sigma = 0$$

3.2.2. *Model B.1/0: (Mixed model with first gradient in D and zero gradient on S)*

The linear functional is expressed in the form

$$\delta\tau = \int\int\int_D \left(F_i\,\zeta^i - \sigma_i^j\,\zeta^i_{,j}\right) dv + \int\int_S T_i\,\zeta^i\,ds$$

Stokes formula yields the separated form

$$\delta\tau = \int\!\!\int\!\!\int_D \left(F_i + \sigma^j_{i,j}\right)\zeta^i\,dv + \int\!\!\int_S \left(T_i - n_j\sigma^j_i\right)\zeta^i\,ds \qquad (S.0)$$

and we deduce the equation of motion in the same form as Eq. (5) and the boundary condition

$$\forall\,\mathbf{x}\in S,\quad n_j\,\sigma^j_i = T_i \;\Leftrightarrow\; \mathbf{n}^\star\sigma = \mathbf{T}$$

Model $B.1/0$ is the classical theory for elastic media and fluids in continuum mechanics.

3.2.3. *Model B.1*

The linear functional is expressed in the form

$$\delta\tau = \int\!\!\int\!\!\int_D \left(F_i\,\zeta^i - \sigma^j_i\,\zeta^i_{,j}\right)dv + \int\!\!\int_S \left(T_i\,\zeta^i + \gamma^j_i\,\zeta^i_{,j}\right)ds \qquad (6)$$

where the tensor γ of components γ^j_i is a new term. The boundary of D is a surface S shared in a partition of N parts S_p of class C^2, $(p = 1, ..., N)$ (Fig. 2). We denote by $(R_m)^{-1}$ the mean curvature of S; the edge Γ_p of S_p is the union of the limit edges Γ_{pq} between surfaces S_p and S_q assumed to be of class C^2 and $\mathbf{t}$ is the tangent vector to Γ_p oriented by $\mathbf{n}$; $\mathbf{n}'$ is the unit external normal vector to Γ_p in the tangent plane to S_p: $\mathbf{n}' = \mathbf{t}\times\mathbf{n}$. Let us notice that:

$$\gamma^j_i\,\zeta^i_{,j} = -\gamma^j_{i,j}\,\zeta^i + V^j_{,j} \qquad (7)$$

where $V^j = \gamma^j_i\,\zeta^i$; consequently, from integration of the divergence of

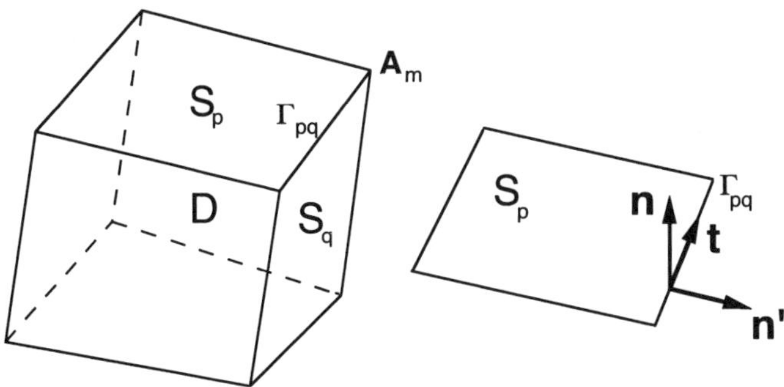

Fig. 2. The set D has a surface boundary S divided in several parts. The edge of S is denoted by Γ which is also divided in several parts with end points $\mathbf{A}_m$.

vector $\mathbf{V}$ on surfaces S_p we obtain,

$$\int\!\!\int_{S_p} V^j_{,j}\, ds = -\int\!\!\int_{S_p} n_j \left(\frac{V^j}{R_m} - V^j_{,l} n^l\right) ds + \int_{\Gamma_p} n'_j V^j\, d\ell \qquad (8)$$

We emphasize with the fact that $V^j_{,l}\, n^l$ corresponds to the normal derivative to S_p denoted $\dfrac{dV^j}{dn}$. An integration by parts of the term $\sigma^j_i\, \zeta^i_{,j}$ in relation (6) and taking account of relations (7-8) implies

$$\delta\tau = \int\!\!\int\!\!\int_D F^1_i \zeta^i\, dv + \int\!\!\int_S T^1_i \zeta^i\, ds + \int\!\!\int_S L_i \frac{d\zeta^i}{dn}\, ds + \sum_{p=1}^N \int_{\Gamma_p} R_{pi}\, \zeta^i d\ell \tag{S.1}$$

with the following definitions

$$\begin{cases} F^1_i \equiv F_i + \sigma^j_{i,j}, & L_i \equiv n_j\, \gamma^j_i \\ T^1_i \equiv T_i - n_j \left(\sigma^j_i - \dfrac{d}{dn}(\gamma^j_i) + \dfrac{1}{R_m}\gamma^j_i\right) - \gamma^j_{i,j}, & R_{pi} \equiv n'_j\, \gamma^j_i \end{cases}$$

Due to theorem 37 in[9] , the distribution $(S.1)$ has a unique decomposition in displacements and transverse derivatives of displacements on the manifolds associated with D and its boundaries: expression $(S.1)$ is in a separated form. Consequently, the equation of motion is

$$\forall\, \mathbf{x} \in D,\ F^1_i = 0 \Leftrightarrow \mathbf{F}^1 = 0$$

and the boundary conditions are

$$\forall\, \mathbf{x} \in S,\ T^1_i = 0, L_i = 0 \Leftrightarrow \mathbf{T}^1 = 0,\ \mathbf{L} = 0$$
$$\forall\, \mathbf{x} \in \Gamma_{pq},\ R_{pi} + R_{qi} = 0 \Leftrightarrow \mathbf{R}_p + \mathbf{R}_q = 0$$

Term $\mathbf{L}$ is not reducible to a force: its virtual work $L_i \dfrac{d\zeta^i}{dn}$ is not the product of a force with the displacement ζ; the term $\mathbf{L}$ is an *embedding action*.

3.3. *Model of second gradient*

3.3.1. *Model A.2*

The linear functional is in the form

$$\delta\tau = \int\!\!\int\!\!\int_D \left(F_i\, \zeta^i - \sigma^j_i\, \zeta^i_{,j} + S^{jk}_i \zeta^i_{,jk}\right) dv$$

Tensor $\mathbf{S}$ with $S^{jk}_i = S^{kj}_i$ is an *overstress tensor*. An integration by parts of the last term brings back to the model $B.1$,

$$\delta\tau = \int\!\!\int\!\!\int_D \left(F_i\, \zeta^i - \left(\sigma^j_i + S^{jk}_{i,k}\right) \zeta^i_{,j}\right) dv + \int\!\!\int_S n_k S^{jk}_i\, \zeta^i_{,j}\, ds$$

and the virtual work gets the separated form $(S.1)$ with:

$$
\begin{cases}
F_i^1 = F_i + \sigma_{i,j}^j + S_{i,jk}^{jk} & \text{volume force} \\
T_i^1 = -n_j \left(\sigma_i^j + S_{i,k}^{jk} - \dfrac{d}{dn} \left(n_k S_i^{jk} \right) + \dfrac{1}{R_m} n_k S_i^{jk} \right) & \text{surface force} \\
R_{pi} = n_j' \, n_k \, S_i^{jk} & \text{line force} \\
L_i = n_j \, n_k \, S_i^{jk} & \text{embedding action}
\end{cases}
$$

and consequently yields the same equation of motion and boundary conditions as in case $B.1$.

3.3.2. Model B.2

The linear functional is in the form

$$
\delta\tau = \int\int\int_D \left(F_i \zeta^i - \sigma_i^j \zeta_{,j}^i + S_i^{jk} \zeta_{,jk}^i \right) dv + \int\int_S \left(T_i \zeta^i + \gamma_i^j \zeta_{,j}^i + U_i^{jk} \zeta_{,jk}^i \right) ds
$$

This functional yields two integrations successively on S_p and on Γ_{pq} with terms at the points A_m. With obvious notations, for the same reasons as in section $3.2.3$, the virtual work gets the separated form

$$
\delta\tau = \int\int\int_D F_i^1 \, \zeta^i dv + \int\int_S \left(T_i^1 \, \zeta^i + L_i^1 \frac{d\zeta^i}{dn} + L_i^2 \frac{d^2\zeta^i}{dn^2} \right) ds
$$
$$
+ \sum_p \int_{\Gamma_p} \left(R_{pi} \, \zeta^i + M_{pi} \frac{d\zeta^i}{dn'} \right) d\ell + \sum_m \phi_{mi} \, \zeta_{A_m}^i \tag{S.2}
$$

where $\zeta_{A_m}^i$ $(i = 1, 2, 3)$ are the components of ζ at point $\mathbf{A}_m$. The calculations are not expended. They introduce the curvature tensor on S_p and the geodesic curvature of Γ_{pq}. Consequently, F_i^1, T_i^1, R_{pi}, ϕ_{mi} are associated with volume, surface, line and forces at points; L_i^1, L_i^2, M_{pi} are embedding efforts of order 1 and 2 on S and of order 1 on the edge Γ. The equation of motion and boundary conditions express that these seven tensorial quantities are null on their domains of values D, S, Γ_p and $\mathbf{A}_m$.

4. Conclusion

It is possible to extend the previous presentation by means of more complex medium with *gradient of order n*. The models introduce embedding effects of more important order on surfaces, edges and points. The *(A.n)* model refers to a *(B.n-1)* model: the fact that boundary surface S is (or is not) a material surface has now a physical meaning. Consequently, we can resume the previous presentation as follows:

a) The choice of a model corresponds to specify the part G of the algebraic dual $T^*_\varphi(\wp)$ in which the efforts are considered: $\Im \in G \subset T^*_\varphi(\wp)$.

b) In order to operate with the principle of virtual works and to obtain the mechanical equations in the form $\Im = 0$, it is no matter that the part G of the dual is separating ($\forall \Im \in G, < \Im, \delta\varphi >= 0 \Rightarrow \delta\varphi = 0$), but it is important the part G is separated ($\Im \in G, \forall \delta\varphi \in T_\varphi(\wp), < \Im, \delta\varphi >= 0 \Rightarrow \Im = 0$).

c) The functionals $A.1$, $B.1/0$, $A.2$, $B.2$ are not separated: if $\Im$ consists in the data of the fields $\mathbf{F}$, σ, $\mathbf{T}$, it is not possible to conclude that the fields are zero.

d) Functionals in $A.0$, $B.0$, $S.1$, $S.2 \ldots$ are separated: if the fields $\mathbf{S}^1$, $\mathbf{T}^1$, $\mathbf{R}^1$, $\mathbf{L}, \ldots$ are continuous then, by using the fundamental lemma of variation calculus, their values must be equal to zero. They are the only functionals we must know for using the principle of virtual works; it is exactly as for a solid: the torque of forces is only known in the equations of motion.

e) When the fields are not continuous on surfaces or curves, we have to consider a model of greater order in gradients and to introduce integrals on inner boundaries of the medium.

For conservative medium, the first gradient theory corresponds to the compressible case. The theory of fluid, elastic, viscous and plastic media refers to the model $(S.0)$. The Laplace theory of capillarity in fluids refers to the model $(S.1)$. To take into account superficial effects acting between solids and fluids, we use the model of fluids endowed with capillarity $(S.2)$; the theory interprets the capillarity in a continuous way and contains the Laplace theory of capillarity; for solids, the model corresponds to "elastic materials with couple stresses" indicated by Toupin in[12] .

5. Example 1: The Laplace theory of capillarity

Liquid-vapor and two-phase interfaces are represented by a material surface endowed with an energy relating to Laplace surface tension. The interface appears as a surface separating two media with its own characteristic behavior and energy properties[13] (when working far from critical conditions, the capillary layer has a thickness equivalent to a few molecular beams[14]). The Laplace theory of capillarity refers to the model $B.1$ in the form $(S.1)$ as following: for a compressible fluid with a capillary effect on the wall boundaries, the free energy is in the form

$$\chi = \int\int\int_D \rho\, \alpha(\rho)\, dv + \sum_{p=1}^N a_p \int\int_{S_p} ds$$

where $\alpha(\rho)$ is the fluid specific energy, ρ is the matter density and coefficients a_p are the surface tensions of each surface S_p. Surface integrations are associated to the space metric; the virtual work of internal forces is

$$\delta\tau_{int} = \int\int\int_D -p_{,i}\,\zeta^i dv + \sum_{p=1}^{N}\int\int_{S_p} n_i\left(p - \frac{a_p}{R_m}\right)\zeta^i\,ds + \sum_{p=1}^{N}\int_{\Gamma_p} a_p n_i'\,\zeta^i\,d\ell$$

where $p \equiv \rho^2\alpha'(\rho)$ is the fluid pressure. The external force (including inertial forces) is the body force $\rho\,\mathbf{f}$ defined in D, the surface force is $\mathbf{T}$ defined on S and the line force is $\mathbf{R}$ defined on Γ. D'Alembert-Lagrange principle yields the equation of motion and boundary conditions:

$$\forall\,\mathbf{x} \in D, \quad -p_{,i} + \rho\,f_i = 0 \quad \Leftrightarrow \quad -\operatorname{grad} p + \rho\,\mathbf{f} = 0,$$

$$\forall\,\mathbf{x} \in S, \quad p\,n_i + T_i - a_p\frac{n_i}{R_m} = 0 \quad \Leftrightarrow \quad p\,\mathbf{n} + \mathbf{T} - a_p\frac{\mathbf{n}}{R_m} = 0$$

$$\forall\,\mathbf{x} \in \Gamma_{pq}, \quad a_p n_{pi}' + a_q n_{qi}' + R_i = 0 \quad \Leftrightarrow \quad a_p\mathbf{n}_p' + a_q\mathbf{n}_q' + \mathbf{R} = 0$$

Boundary conditions are *Laplace equation* and *Young-Dupré condition*.

6. Example 2: Fluids endowed with internal capillarity

For interfacial layers, kinetic theory of gas leads to laws of state associated with non-convex internal energies[15,16]. This approach dates back to van der Waals[17], Korteweg[18], corresponds to the Landau-Ginzburg theory[19] and presents two disadvantages. First, between phases, the pressure may become negative; simple physical experiments can be used to cause traction that leads to these negative pressure values[20]. Second, in the field between bulks, internal energy cannot be represented by a convex surface associated with density and entropy; this fact seems to contradict the existence of equilibrium states; it is possible to eliminate this disadvantage by writing in an anisotropic form the stress tensor of the capillary layer which allows to study interfaces of non-molecular size near a critical point.
One of the problems that complicates this study of phase transformation dynamics is the apparent contradiction between Korteweg classical stress theory and the Clausius-Duhem inequality[21]. Proposal made by Eglit[22], Dunn and Serrin[23], Casal and Gouin[24] and others rectifies this anomaly for liquid-vapor interfaces. The simplest model in continuum mechanics considers a free energy as the sum of two terms: a first one corresponding to a medium with a uniform composition equal to the local one and a second one associated with the non-uniformity of the fluid[15,17]. The second term is approximated by a gradient expansion, typically truncated to

the second order. The model is simpler than models associated with the renormalization-group theory[25] but has the advantage of easily extending well-known results for equilibrium cases to the dynamics of interfaces[26–28]. We consider a fluid D in contact with a wall S. Physical experiments prove that the fluid is nonhomogeneous in the neighborhood of S [16]. The internal energy ε is also a function of the entropy. In the case of isothermal motions, the internal energy is replaced by the free energy. In the mechanical case, the entropy and the temperature are not concerned by the virtual displacements of the medium. Consequently, for isentropic or isothermal motions, $\varepsilon = f(\rho, \beta)$ where $\beta = (\operatorname{grad} \rho)^2$. The fluid is submitted to external forces represented by a potential Ω as a function of Eulerian variables $\mathbf{x}$. To obtain boundary conditions it is necessary to know the wall effect. An explicit form for the energy of interaction between surfaces and liquids is proposed in[29]. We denote by B the surface density of energy at the wall. The total energy E of the fluid is the sum of three potential energies: E_f (bulk energy), E_p (external energy) and E_S (surface energy).

$$E_f = \iiint_D \rho\,\varepsilon(\rho,\beta)\,dv, \quad E_p = \iiint_D \rho\,\Omega(\mathbf{x})\,dv, \quad E_S = \iint_S B\,ds$$

We have the results (see Appendix): $\delta E_f = \iiint_D (-\operatorname{div}\sigma)\,\zeta\,dv$

$$+ \iint_S \left\{ -A\frac{d\zeta_n}{dn} + \left(\frac{2A}{R_m}\mathbf{n}^* + \operatorname{grad}^*_{tg}A + \mathbf{n}^*\sigma\right)\zeta \right\}ds - \int_\Gamma A\,\mathbf{n}'^*\,\zeta\,d\ell$$

with $\sigma = -P\mathbf{1} - C\,\operatorname{grad}\rho \otimes \operatorname{grad}\rho \equiv -P\mathbf{1} - C\left(\frac{\partial\rho}{\partial\mathbf{x}}\right)^*\frac{\partial\rho}{\partial\mathbf{x}}$, where $C = 2\rho\,\varepsilon'_\beta$, $P = \rho^2\varepsilon'_\rho - \rho\,\operatorname{div}(C\,\operatorname{grad}\rho)$, ε'_ρ (or ε'_β) denoting the partial derivative of ε with respect to ρ (or β), $\zeta_n = \mathbf{n}^*\zeta$; $A = C\rho\,\dfrac{d\rho}{dn}$ where $\dfrac{d\rho}{dn} = \dfrac{\partial\rho}{\partial\mathbf{x}}\mathbf{n}$ and grad_{tg} denotes the tangential part of the gradient relatively to S.

$$\delta E_p = \iiint_D \rho\,\frac{\partial\Omega}{\partial\mathbf{x}}\,\zeta\,dv \equiv \iiint_D \rho\,(\operatorname{grad}^*\Omega)\,\zeta\,dv; \quad \text{and,}$$

$$\delta E_S = \iint_S \left\{ \delta B - \left(\frac{2B}{R_m}\mathbf{n}^* + \operatorname{grad}^*_{tg}B\right)\zeta \right\}ds + \int_\Gamma B\,\mathbf{n}'^*\,\zeta\,d\ell$$

The density in the fluid has a limit value ρ_s at the wall S and B is assumed to be a function of ρ_s only[29]. Then, $\delta B = B'(\rho_s)\,\delta\rho_s = -\rho_s B'(\rho_s)\,\operatorname{div}\zeta$, where $\operatorname{div}\zeta$ is computed on S [11]. Let us denote $G = -\rho_s B'(\rho_s)$; Appendix yields

$$\iint_S \delta B\,ds = \iint_S G\,\operatorname{div}\zeta\,ds =$$

$$= \int\int_S \left\{ G\frac{d\zeta_n}{dn} - \left(\frac{2G}{R_m}\,\mathbf{n}^* + \mathrm{grad}_{tg}^* G \right) \zeta \right\} ds + \int_\Gamma G\,\mathbf{n}'^*\,\zeta\,d\ell$$

$$\delta E_S = \int\int_S \left\{ G\frac{d\zeta_n}{dn} - \left(\frac{2H}{R_m}\,\mathbf{n}^* + \mathrm{grad}_{tg}^* H \right) \zeta \right\} ds + \int_\Gamma H\,\mathbf{n}'^*\,\zeta\,d\ell$$

with $H = B(\rho_S) - \rho_S B'(\rho_S)$. Then,

$$\begin{aligned}
\delta E = {} & \int\int\int_D (\rho\,\mathrm{grad}^*\,\Omega - \mathrm{div}\,\sigma)\,\zeta\,dv \; - \int_\Gamma (A - H)\,\mathbf{n}'^*\,\zeta\,d\ell \\
& + \int\int_S (G - A)\frac{d\zeta_n}{dn} + \left(\frac{2(A - H)}{R_m}\,\mathbf{n}^* + \mathrm{grad}_{tg}^*(A - H) + \mathbf{n}^*\sigma \right) \zeta\,ds
\end{aligned} \tag{9}$$

At equilibrium, $\delta\tau \equiv -\delta E = 0$. The fundamental lemma of variation calculus associated with separated form (9) corresponding to $(S.2)$, yields:

6.1. *Equation of equilibrium*

From any arbitrary variation $\mathbf{x} \in D \to \zeta(\mathbf{x})$ such that $\zeta = \mathbf{0}$ on S, we get

$$\int\int\int_D (\rho\,\mathrm{grad}^*\Omega - \mathrm{div}\,\sigma)\,\zeta\,ds = 0. \quad \text{Then,}$$

$$\rho\,\mathrm{grad}^*\Omega - \mathrm{div}\,\sigma = 0 \tag{10}$$

This equation is written in the classical form of equation of equilibrium[24] . *It is not the same for the boundary conditions.*

6.2. *Boundary conditions*

6.2.1. *Case of a rigid wall*

We consider a rigid wall; on S, the virtual displacements satisfy the condition $\mathbf{n}^* \zeta = 0$. Then, at the rigid wall, $\forall\,\mathbf{x} \in S \to \zeta(\mathbf{x})$ such that $\mathbf{n}^* \zeta = 0$,

$$\int\int_S (G - A)\frac{d\zeta_n}{dn} + \left(\frac{2(A - H)}{R_m}\,\mathbf{n}^* + \mathrm{grad}_{tg}^*(A - H) + \mathbf{n}^*\sigma \right) \zeta\,ds = 0$$

Due to $\sigma = \sigma^*$, we deduce the boundary conditions (11-12)

$$\forall\,\mathbf{x} \in S,\ G - A = 0, \tag{11}$$

and there exists a Lagrange multiplier $\mathbf{x} \in S \to \lambda(\mathbf{x}) \in R$ such that,

$$\forall\,\mathbf{x} \in S,\ \frac{2(A - H)}{R_m}\,\mathbf{n} + \mathrm{grad}_{tg}(A - H) + \sigma\,\mathbf{n} = \lambda\,\mathbf{n} \tag{12}$$

The edge Γ of S belongs to the solid wall and consequently on Γ, $\zeta = \eta\,\mathbf{t}$: the integral on Γ is null and does not yield any additive condition.

6.2.2. *Case of an elastic wall*

The equilibrium equation (10) is unchanged. On S, the condition (11) is also unchanged. The only different condition comes from the fact that we do not have anymore the slipping condition for the virtual displacement on S, $(\mathbf{n}^* \boldsymbol{\zeta} = 0)$. Due to the possible deformation of the wall, the virtual work of stresses on S is $\delta E_e = \displaystyle\int\!\!\int_S \boldsymbol{\kappa}^* \boldsymbol{\zeta} \, ds + \int_\Gamma \mathbf{R}^* \boldsymbol{\zeta} \, d\ell$ where $\boldsymbol{\kappa} = \sigma_e \, \mathbf{n}$ is the stress (loading) vector associated with stress tensor σ_e of the elastic wall and $\mathbf{R}$ is the line force due to the elasticity of the line. Relation (12) is replaced by

$$\forall \, \mathbf{x} \in S, \; 2 \, \frac{(A - H)}{R_m} \, \mathbf{n} + \mathrm{grad}_{tg}(A - H) + \sigma \, \mathbf{n} + \boldsymbol{\kappa} = 0$$

We obtain an additive condition on Γ in the form $(H - A) \, \mathbf{n}' + \mathbf{R} = 0$ and due to condition (11),

$$\forall \, \mathbf{x} \in \Gamma, \; B \, \mathbf{n}' + \mathbf{R} = 0 \tag{13}$$

(If Γ is the union of edges Γ_p, $B \, \mathbf{n}'$ is replaced by $\displaystyle\sum_p B_p \, \mathbf{n}'_p$).

6.3. *Analysis of the boundary conditions*

Eq. (11) yields $\; C \, \dfrac{d\rho}{dn} + B'(\rho_s) = 0$; the definition of σ implies:

$$\sigma \, \mathbf{n} = -P\mathbf{n} - C \, \frac{d\rho}{dn} \, \mathrm{grad} \, \rho$$

Due to the fact that the tangential part of Eq. (12) is always verified, the only condition comes from Eq. (11); Eq. (12) yields the value of the Lagrange multiplier λ and Eq. (13) the value of $\mathbf{R}$. For an elastic (non-rigid) wall we obtain,

$$\kappa_{tg} = 0 \quad \text{and} \quad \kappa_n = P + \frac{2B}{R_m} - B'(\rho_s) \, \frac{d\rho}{dn} \tag{14}$$

where κ_{tg} and κ_n are the tangential and the normal components of $\boldsymbol{\kappa}$. Taking into account of Eq. (14) we obtain the stress values at the non rigid elastic wall. The surface energy is[29] : $B(\rho_s) = -\gamma_1 \, \rho_s + \dfrac{\gamma_2}{2} \, \rho_s^2$ where γ_1 and γ_2 are two positive constants and the fluid density condition at the wall is

$$C \, \frac{d\rho}{dn} = \gamma_1 - \gamma_2 \, \rho_s$$

If we denote by $\rho_B = \gamma_1/\gamma_2$ the *bifurcation fluid density* at the wall, due to the fact C is positive constant[16], we obtain: if $\rho_S < \rho_B$, (or $\rho_S > \rho_B$), $\dfrac{d\rho}{dn}$ is positive (or negative) and we have a lack (or excess) of fluid density at the wall. Such media allow to study fluid interfaces and interfacial layers between fluids and solids and lead to numerical and asymptotic methods[30]. The extension to the dynamic case is straightforward: Eq. (10) yields

$$\rho\,\mathbf{\Gamma}^* - \operatorname{div}\sigma + \rho\,\operatorname{grad}^*\Omega = 0$$

Vector $\mathbf{\Gamma}$ is the acceleration; boundary conditions (11-14) are unchanged.

Acknowledgments

I am grateful to Professor Tommaso Ruggeri for helpful discussions.

Appendix

Let S be a surface in the 3-dimensional space and $\mathbf{n}$ its external normal extended locally in the vicinity of S by the expression $\mathbf{n}(\mathbf{x}) = \operatorname{grad} d(\mathbf{x})$, where d is the distance of a point $\mathbf{x}$ to S; for any vector field $\mathbf{w}$, we obtain[31] :

$$\operatorname{rot}(\mathbf{n}\times\mathbf{w}) = \mathbf{n}\operatorname{div}\mathbf{w} - \mathbf{w}\operatorname{div}\mathbf{n} + \frac{\partial\mathbf{n}}{\partial\mathbf{x}}\mathbf{w} - \frac{\partial\mathbf{w}}{\partial\mathbf{x}}\mathbf{n}$$

From $\mathbf{n}^*\dfrac{\partial\mathbf{n}}{\partial\mathbf{x}} = 0$ and $\operatorname{div}\mathbf{n} = -\dfrac{2}{R_m}$ we deduce on S,

$$\mathbf{n}^*\operatorname{rot}(\mathbf{n}\times\mathbf{w}) = \operatorname{div}\mathbf{w} + \frac{2}{R_m}\mathbf{n}^*\mathbf{w} - \mathbf{n}^*\frac{\partial\mathbf{w}}{\partial\mathbf{x}}\mathbf{n} \qquad (15)$$

We deduce: for any scalar field A and $\mathbf{w} = A\boldsymbol{\zeta}$,

$$A\operatorname{div}\boldsymbol{\zeta} = A\frac{d\zeta_n}{dn} - \frac{2A}{R_m}\zeta_n - (\operatorname{grad}^*_{tg}A)\boldsymbol{\zeta} + \mathbf{n}^*\operatorname{rot}(A\mathbf{n}\times\boldsymbol{\zeta}) \qquad (16)$$

Let us calculate δE_f; D is a material volume, then $\delta E_f = \displaystyle\int\!\!\int\!\!\int_D \rho\,\delta\varepsilon\,dv$ with $\delta\varepsilon = \dfrac{\partial\varepsilon}{\partial\rho}\,\delta\rho + \dfrac{\partial\varepsilon}{\partial\beta}\,\delta\beta$. From $\delta\dfrac{\partial\rho}{\partial\mathbf{x}} = \dfrac{\partial\delta\rho}{\partial\mathbf{x}} - \dfrac{\partial\rho}{\partial\mathbf{x}}\dfrac{\partial\boldsymbol{\zeta}}{\partial\mathbf{x}}$ (see [24]),

$$\rho\,\varepsilon'_\beta\,\delta\beta = 2\rho\,\varepsilon'_\beta\,\delta\frac{\partial\rho}{\partial\mathbf{x}}\left(\frac{\partial\rho}{\partial\mathbf{x}}\right)^* = C\left(\frac{\partial\delta\rho}{\partial\mathbf{x}} - \frac{\partial\rho}{\partial\mathbf{x}}\frac{\partial\boldsymbol{\zeta}}{\partial\mathbf{x}}\right)\left(\frac{\partial\rho}{\partial\mathbf{x}}\right)^*$$

$$= \operatorname{div}(C\operatorname{grad}\rho\,\delta\rho) - \operatorname{div}(C\operatorname{grad}\rho)\,\delta\rho - tr\left(C\operatorname{grad}\rho\operatorname{grad}^*\rho\,\frac{\partial\boldsymbol{\zeta}}{\partial\mathbf{x}}\right)$$

Due to $\delta\rho = -\rho \operatorname{div}\zeta$ (see [11]),

$$\delta E_f = \iiint_D \left(\frac{\partial P}{\partial \mathbf{x}} + \operatorname{div}(C \operatorname{grad}\rho \operatorname{grad}^*\rho)\right)\zeta \, dv$$

$$- \iiint_D \operatorname{div}\left(C\rho \operatorname{grad}\rho \operatorname{div}\zeta + C \operatorname{grad}\rho \operatorname{grad}^*\rho\,\zeta + P\,\zeta\right) dv$$

$$= \iiint_D -(\operatorname{div}\sigma)\,\zeta\, dv + \iint_S (-A \operatorname{div}\zeta + \mathbf{n}^*\sigma\,\zeta)ds$$

From Eq. (16), we deduce immediatly: $\quad \delta E_f = \iiint_D (-\operatorname{div}\sigma)\zeta\, dv \; +$

$$\iint_S \left\{-A\frac{d\zeta_n}{dn} + \left(\frac{2A}{R_m}\,\mathbf{n}^* + \operatorname{grad}^*_{tg} A + \mathbf{n}^*\sigma\right)\zeta\right\} ds - \int_\Gamma A\,\mathbf{n}'^*\,\zeta\, d\ell$$

Let us calculate δE_S; due to $\quad E_S = \iint_S B \, \det(\mathbf{n}, d_1\mathbf{x}, d_2\mathbf{x})\quad$ where $d_1\mathbf{x}$ and $d_2\mathbf{x}$ are two coordinate lines of S, we get:

$$E_S = \iint_{S_0} B \, \det F \, \det(F^{-1}\mathbf{n}, d_1\mathbf{X}, d_2\mathbf{X})$$

where S_0 is the image of S in a reference space with Lagrangian coordinates $\mathbf{X}$ and F is the deformation gradient tensor $\dfrac{\partial \mathbf{x}}{\partial \mathbf{X}}$ of components $\left\{\dfrac{\partial x^i}{\partial X^j}\right\}$.

Then, $\quad \delta E_S = \iint_{S_0} \delta B \, \det F \, \det(F^{-1}\mathbf{n}, d_1\mathbf{X}, d_2\mathbf{X})$

$$+ \iint_{S_0} B\,\delta\left(\det F \, \det(F^{-1}\mathbf{n}, d_1\mathbf{X}, d_2\mathbf{X})\right).$$

with $\quad \displaystyle\iint_{S_0} B\,\delta\left(\det F \, \det(F^{-1}\mathbf{n}, d_1\mathbf{X}, d_2\mathbf{X})\right) =$

$$\iint_S B \operatorname{div}\zeta \, \det(\mathbf{n}, d_1\mathbf{x}, d_2\mathbf{x}) + B \det\left(\frac{\partial \mathbf{n}}{\partial \mathbf{x}}\,\zeta, d_1\mathbf{x}, d_2\mathbf{x}\right)$$

$$-B\det\left(\frac{\partial\zeta}{\partial\mathbf{x}}\,\mathbf{n}, d_1\mathbf{x}, d_2\mathbf{x}\right) = \iint_S \left(\operatorname{div}(B\,\zeta) - (\operatorname{grad}^*B)\,\zeta - B\mathbf{n}^*\frac{\partial\zeta}{\partial\mathbf{x}}\,\mathbf{n}\right) ds$$

Relation (15) yields: $\operatorname{div}(B\,\zeta) + \dfrac{2B}{R_m}\,\mathbf{n}^*\zeta - \mathbf{n}^*\dfrac{\partial B\,\zeta}{\partial\mathbf{x}}\,\mathbf{n} = \mathbf{n}^* \operatorname{rot}(B\,\mathbf{n}\times\zeta)$,

$$\iint_{S_0} B\,\delta\left(\det F \, \det(F^{-1}\mathbf{n}, d_1\mathbf{X}, d_2\mathbf{X})\right) =$$

$$\iint_{S_0} \left(-\frac{2B}{R_m}\,\mathbf{n}^* + \operatorname{grad}^*B\,(\mathbf{n}\mathbf{n}^* - \mathbf{1})\right)\zeta\, ds + \iint_S \mathbf{n}^* \operatorname{rot}(B\,\mathbf{n}\times\zeta)\, ds$$

where $\operatorname{grad}^*B\,(\mathbf{n}\mathbf{n}^* - \mathbf{1})$ belongs to the cotangent plane to S; we obtain

$$\delta E_S = \iint_S \left(\delta B - \left(\frac{2B}{R_m}\,\mathbf{n}^* + \operatorname{grad}^*_{tg}B\right)\zeta\right) ds + \int_\Gamma B\,\mathbf{n}'^*\zeta\, d\ell.$$

References

1. P. Germain, *Cours de mécanique des milieux continus* (Masson, Paris, 1973).
2. C. Truesdell, *First course in rational continuum mechanics* (Academic Press, New York, 1994).
3. E. Cosserat and F. Cosserat, *Sur la théorie des corps déformables* (Hermann, Paris, 1909).
4. M.E. Gurtin, *Configurational forces as basic concepts in continuum physics* (Springer, Berlin, 2000).
5. W. Noll and E.G. Virga, *Arch. Rat. Mech. Anal.* **3**, 1 (1990).
6. F. dell'Isola and P. Seppecher, *Meccanica* **32**, 33 (1997).
7. P. Germain, *J. Mécanique*, **12**, 235 (1973).
8. P. Germain, *S.I.A.M. J. Appl. Math.* **25**, 556 (1973).
9. L. Schwartz, *Théorie des distributions*, Ch. 3 (Hermann, Paris, 1966).
10. H. Gouin and F. Gouin, *Mech. Res. Comm.* **10**, 21 (1983).
11. J. Serrin, Mathematical principles of classical fluid mechanics, in *Encyclopedia of Physics VIII/1*, Ed: S. Flügge, (Springer, Berlin, 1960) pp. 125-263.
12. R.A. Toupin, *Arch. Rat. mech. Anal.* **11**, 385 (1962).
13. V. Levitch, *Physicochemical Hydrodynamics* (Prentice-Hall, New Jersey, 1962).
14. S. Ono and S. Kondo, Molecular theory of surface tension in liquid, in *Encyclopedia of Physics, X*, Ed: S. Flügge (Springer, Berlin, 1960).
15. J.W. Cahn and J.E. Hilliard, *J. Chem. Phys.* **31**, 688 (1959).
16. J.S. Rowlinson. and B. Widom, *Molecular theory of capillarity* (Clarendon Press, Oxford, 1984).
17. J.D. van der Waals, *Archives Néerlandaises* **28**, 121 (1894-1895).
18. J. Korteweg, *Archives Néerlandaises* **2**, **n° 6**, 1 (1901).
19. P.C. Hohenberg and B.I. Halperin, *Rev. Mod. Phys.* **49**, 435 (1977).
20. Y. Rocard, *Thermodynamique* (Masson, Paris, 1952).
21. M.E. Gurtin, *Arch. Rat. Mech. Anal.* **19**, 339 (1965).
22. M.E. Eglit, *J. Appl. Math. Mech.* **29**, 351 (1965).
23. J.E. Dunn and J. Serrin, *Arch. Rat. Mech. Anal.* **88**, 95 (1985).
24. P. Casal and H. Gouin, *C. R. Acad. Sci. Paris* **300, II**, 231 (1985).
25. C. Domb, *The critical point* (Taylor & Francis, London, 1996).
26. M. Slemrod, *Arch. Rat. Mech. Anal.* **81**, 301 (1983).
27. L. Truskinovsky, *P.M.M.* **51**, 777 (1987).
28. H. Gouin and T. Ruggeri, *Eur. J. Mech B/ Fluids* **24**, 596 (2005).
29. H. Gouin, *J. Phys. Chem. B* **102**, 1212 (1998).
30. H. Gouin and S.L. Gavrilyuk, *Physica A*, **268**, 291 (1999).
31. S. Kobayashi and K. Nomizu, *Foundations of differential geometry*, vol. 1 (Interscience Publ., New York, 1963).

A MODEL FOR THE EVOLUTION OF BIOENERGY IN AN ENVIROMENTAL SYSTEM

G. Lauro

Facoltà di Architettura, II Università di Napoli, Italy
E-mail: giuliana.lauro@unina2.it

R. Monaco, G. Servente

Dipartimento di Matematica, Politecnico di Torino, Italy
E-mail: roberto.monaco@polito.it, giorgiaservente@yahoo.com

A dynamical system representing the time evolution of bioenergy in an environmental system is here proposed. After the derivation of the model and a detailed description of the parameters, the equilibria of the dynamical system are detected and the relative stability analysis is performed. Finally an application to a real case is shown.

1. The mathematical model

In the modern discipline of Landscape Ecology,[4,5,7] the landscape is defined as a heterogenous land composed of interacting ecosystems that exchange energy and matter and where natural and anthropic events coexist. The study of the dynamics of such complex ecological system requires the use of models that simulate how it operates under the action of constraints imposed by environmental limits.

In this paper we consider an environmental system splitted in several different ecological patches separated from each other by natural or anthropic barriers. These barriers, as we will see further on, can have different degrees of permeability to the migration of bioenergy.
Such a system can be represented by Fig.1, where we have assumed that the system is divided into nine different ecological patches.
These patches, correlated by more or less fast energy exchanges,[7] correspond to different geographical areas characterized by peculiar biotic and abiotic features.
Thanks to the biological processes developed inside each patch, a different

amount of bioenergy is there produced. This bioenergy must be intended in a generalized sense, since it is given by the product of the patch biopotentiality[7] with a factor which takes into account the morphological features of the patch itself. In other words the bioenergy treated in this work is affected not simply by the actual energy produced by the patch but also by the peculiar ecological features of the areas inside the patch itself.

The bioenergy magnitude of each patch may be represented by a circle (node) whose diameter is proportional to the magnitude itself (see Fig.1). In some cases there can be patches with no bioenergy at all, since they are biologically non-active. The energy exchange among them will be more or less strong depending on the degree of permeability of the barriers which can obstruct the energy passage. Therefore, each patch is connected to the others by links (arches) whose width is proportional to the energy flux shared among them (see again Fig.1). Connectivity between patches can be measured by a connectivity parameter[4] of the environmental system. Detailed computation of bioenergy, biopotentiality, arch widths and connectivity parameter will be given in next section. The collection of nodes and arches is frequently called graph of the environmental system.[4]

The graph so obtained, represents in a static way the exchange of energy that occurs in the territory. Very recently, in the book,[9] models to investigate the time evolution of bioenergy and of other quantities, describing the territory, have been proposed in order to get a dynamic evaluation of the ecological value of the environmental system, starting from its present territorial settlement. The primary objective of these models is not to perform quantitative predictions, but to estimate the goodness of the territorial plan, finding eventually critical values of the quantities characterizing the territory itself.

The model basic assumption[9] is that the time evolution of bioenergy will depend on two terms with opposite signs. The first, positive, describes the bioenergy growth following a logistic law, and it is expressed by $c\left[1 - M(t)/M_{\max}\right]M(t)$, where c is the connectivity parameter and $M_{\max}$ is the maximum value of bioenergy the environmental system can provide. The second term is opposite to bioenergy growth and it is given by $-hS(t)M_{\max}$, where h is the ratio between the sum of the impermeable barrier lengths, inside the environmental system, and its external perimeter. In order to take globally into account what is opposed to bioenergy growth, the quantity h will be multiplied by the scale factor $M_{\max}$ and by the quantity $S(t)$, which is the ratio between the sum of the territory surfaces that, at time t, present low values of biopotentiality and the total

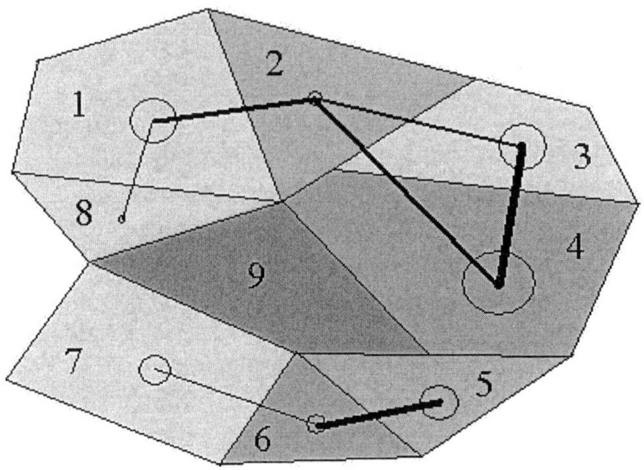

Fig. 1. Graph of the energy exchanges among patches

surface of the system.

Thus, by introducing the normalized bioenergy $\mathcal{M}(t) = M(t)/M_{\mathrm{max}}$ the first equation of the model will be given by

$$\mathcal{M}'(t) = c\mathcal{M}(t)[1 - \mathcal{M}(t)] - hS(t). \tag{1}$$

To complete the derivation of the model it is necessary to derive an equation for the quantity $S(t)$. As we said earlier, this represents the amount of areas with low biopotentiality in the territory under investigation. If we introduce the variable $\mathcal{V}(t)$, that is the ratio between the total amount of territory surfaces characterized by high values of biopotentiality and the whole surface of the system, then we can set $S(t) = 1 - \mathcal{V}(t)$.

In the book[9] an evolution model for $\mathcal{V}(t)$ was proposed and tested. Such a model has the following form

$$\mathcal{V}'(t) = \mathcal{M}(t)\mathcal{V}(t)[1 - \mathcal{V}(t)] - h_R U_o \mathcal{V}(t), \tag{2}$$

where U_o is the constant ratio between the surface of the edified areas (biologically non-active) and the total surface of the environmental system; h_R

is an environment impact parameter that can be defined as the ratio between the sum of the perimeters of the edified areas and the total perimeter of the system.

We can notice that the last equation has again a non-negative logistic term, pre-multiplied by $\mathcal{M}(t)$, and a negative one contrasting the high biopotentiality surfaces expansion.

Equations (2) and (1), that we rewrite setting $S(t) = 1 - \mathcal{V}(t)$, i.e.

$$\mathcal{M}'(t) = c\mathcal{M}(t)[1 - \mathcal{M}(t)] - h[1 - \mathcal{V}(t)], \tag{3}$$

represent the mathematical model to be investigated starting from the initial data $\mathcal{V}(0) = \mathcal{V}_0$ e $\mathcal{M}(0) = \mathcal{M}_0$.

Let us finally observe that the state variables $\mathcal{V}$ and $\mathcal{M}$ and the quantities U_o and c have values in the interval $[0,1]$. Conversely, the parameters h and h_R can take values larger than one, marking, in this case, that the territorial settlement presents some criticality.

2. Computation of parameters and initial data of the model

In this section we will show how to determine the model state variables $\mathcal{M}$ e $\mathcal{V}$, as well as the other parameters of equations (2) and (3).

Let us assume, therefore, to consider an environmental system splitted into m ecological patches.

The bioenergy of the patch j, $j = 1, \ldots, m$, is given by

$$M_j = (1 + K_j)B_j,$$

where $K_j \in [0, 1]$ is a dimensionless parameter characterizing the morphological features of the patch and B_j is the biopotentiality of the patch itself. The last quantity assumes values between 0 and approximately 5 Mcal per year and can be computed, on the basis of a standard classification,[6] once is determined the kind of vegetation present in the area. Such a classification is based on five classes, from the lower concerning edified (biologically non-active) areas to the fifth characterized by natural wooden areas.

The parameter K_j is computed as the average[4] between three parameters K_j^F, K_j^P, K_j^D, each with values in $[0, 1]$.

The first is a parameter related to the shape of the patch borders, since their morphology influences strongly the energy exchanges between the patches themselves.

Conversely, the second parameter K_j^P, again with the purpose of evaluating energy exchanges, takes into account the permeability of the barriers, following some standard values of the permeability parameter that can be found in the book.[4]

Finally, the third parameter K_j^D is related to biodiversity and is determined by a Shannon entropy value,[2] since diversity of ecological areas inside each patch influences once more bioenergy exchanges. Detailed computations of these last three parameters can be found in the book.[9]

Once the bioenergy of each patch is computed, that of the whole system can be determined by performing the average over all the patches. In particular the normalized state variable $\mathcal{M}$ will be given by

$$\mathcal{M} = \frac{1}{mM_{\max}} \sum_{j=1}^{m} M_j, \tag{4}$$

where

$$M_{\max} = 2B_{\max}, \qquad B_{\max} = \max_{j=1,\ldots,m} \{B_j\}.$$

Thus, as we shall see in the last section, from the data that can be recovered from the Geographical Information System (GIS) of the territory under investigation, we can compute the quantity (4) and assume it as the initial datum $\mathcal{M}_0$ for equation (3). Analogously, it is possible from the GIS and the peculiar vegetation inside each patch, to select the areas with high values of biopotentiality and compute the normalized quantity $\mathcal{V}$ to be assumed as initial data for equation (2).

Again from the actual data of the region under investigation it is possible to evaluate the other parameters h, h_R e U_o of the model, already defined. However, the parameter h_R may have a more complicate definition than the one given in the previous section, since it represents the environmental pressure of the low biopotentiality areas on the high ones, due to the presence of pollution agents. In this case h_R can be computed by means of a combination of several environmental variables to be processed by a statistical multivariate analysis;[9] however, for the aims of the present paper, it is sufficient the definition given in the previous section.

For what concerns the connectivity parameter c, first of all one has to compute the bioenergy fluxes through the border between two patches i and j. These are given by[4]

$$F_{ij} = \frac{M_i + M_j}{2} \frac{L_{ij}\, p_{ij}}{P_i + P_j} \tag{5}$$

where L_{ij} is the length of the border, P_i e P_j the perimeters of the two patches and p_{ij} the permeability parameter of the barrier whose value, as already said, is known in literature[4] and depends on the type of the barrier itself.

The number Λ of fluxes F_{ij} depends on the geographical settlement of the territory. Renumbering the fluxes by F^k, $k = 1, \ldots, \Lambda$, the connectivity parameter can be computed through the following weighted average[4]

$$c = \frac{1}{\Lambda} \sum_{k=1}^{\Lambda} \frac{F^k}{\max_k \{F^k\}}. \tag{6}$$

3. Stability analysis of equilibrium solutions

Changes in bioenergy and environmental conditions may produce territorial modifications toward which individual landscapes will tend to move smoothly (attractors) or may produce, instead, critical thresholds that result in radical changes in the state of the ecological system. In this sense, ecological systems are metastable.[8] Stability means that an ecological system remains relatively unchanged and return to the same attractor if subjected to some disturbances. Metastability means that it can maintain itself over a limited range of changes in environmental conditions but may eventually undergo significant alterations if environmental constraints continue to change. As remarked in,[4,6] the more or less metastability (i.e., more or less resistance to disturbances) is related to the more or less presence of biodiversity and connectivity. Hence, it is interesting to study the equilibrium solutions of equations (2)–(3) because they can give, together with the parameters involved, indications on the level of metastability of the model.

By setting the right hand side of (2)–(3) equal to zero, we deal, firstly, with equilibrium solutions:

$$\mathcal{V}_e^1 = 0 \qquad \mathcal{M}_e^1 = \frac{1}{2} \left[1 - \sqrt{1 - 4h/c} \right] \tag{7}$$

$$\mathcal{V}_e^2 = 0 \qquad \mathcal{M}_e^2 = \frac{1}{2} \left[1 + \sqrt{1 - 4h/c} \right]. \tag{8}$$

The existence in the real field of such solutions (both positive) requires to be $h < c/4$. This equilibrium configuration corresponds to a territorial settlement with a lack of areas at high value of biopotentiality ($\mathcal{V} = 0$) but, the condition of low impermeability h and/or high connectivity c, allows to have $\mathcal{M}$ different from zero.

In order to find, instead, equilibrium solutions with $\mathcal{V}_e$ different from zero, we must calculate the solutions of the following third order equation

$$\mathcal{M}^3 - \mathcal{M}^2 + H = 0, \qquad H = hh_R U_o/c. \tag{9}$$

The corresponding values of $\mathcal{V}_e$ are:

$$\mathcal{V}_e = 1 - h_R U_o/\mathcal{M}_e. \tag{10}$$

In order to have positive values of (10) we need that

$$\mathcal{M}_e > h_R U_o. \tag{11}$$

Moreover, from the theory of third order equations,[1] equation (9) will admit a unique real solution if $H > 4/27$, namely:

$$\mathcal{M}_e^3 = \frac{1}{3} + \left[\frac{1}{27} - \frac{H}{2} + \left(\frac{H^2}{4} - \frac{H}{27}\right)^{1/2}\right]^{\frac{1}{3}} + \left[\frac{1}{27} - \frac{H}{2} - \left(\frac{H^2}{4} - \frac{H}{27}\right)^{1/2}\right]^{\frac{1}{3}}.$$

From the definition of parameter H follows that the larger is its value the worse the environmental conditions will be. In fact, to high values of H correspond high values of h, h_R e U_o and/or low values of connectivity c, namely, this case corresponds to an ecological system characterized by a poor flux of energy. Moreover, numerical calculations, made with realistic values of parameter H, have shown a trend of the solution $\mathcal{M}_e^3$ toward negative values, hence toward a not eco-sustainable territorial settlement.

On the contrary, for $H \leq 4/27$, we get three real solutions,[1] one of which is negative, say $\mathcal{M}_e^4$, whereas the other two, say $\mathcal{M}_e^5 < \mathcal{M}_e^6$, are positive, as one can verify by taking into account the range of realistic values of parameters h, h_R, U_o and c.

In order to study the stability of the equilibrium solutions of (2)–(3) we need to linearize such nonlinear system around these solutions $(\mathcal{V}_e^k, \mathcal{M}_e^k)$, $k = 1, \ldots, 6$, and evaluate the eigenvalues $\lambda_{1,2}$ of the relative jacobian matrix:[3]

$$\begin{pmatrix} \mathcal{M}_e^k - 2\mathcal{M}_e^k \mathcal{V}_e^k - h_R U_o & \mathcal{V}_e^k(1 - \mathcal{V}_e^k) \\ h & c - 2c\mathcal{M}_e^k \end{pmatrix} \tag{12}$$

and hence,

$$\lambda_{1,2} = \frac{1}{2}\left[A + B \pm \sqrt{(A+B)^2 - 4(AB - K)}\right], \tag{13}$$

where

$$A = \mathcal{M}_e^k(1 - 2\mathcal{V}_e^k) - h_R U_o, \quad B = c(1 - 2\mathcal{M}_e^k), \quad K = h\mathcal{V}_e^k(1 - \mathcal{V}_e^k).$$

We now analyze the different situations. The equilibrium solutions $(0, \mathcal{M}_e^1)$ and $(0, \mathcal{M}_e^2)$ will be asymptotically stable (negative eigenvalues) if

$$\frac{1}{2} < M_e^{1,2} < h_R U_o. \tag{14}$$

That is, even if the the system is characterized by low biopotentiality it shows a stable response to disturbances. It is clear that the condition (14) excludes the existence of equilibrium solutions $(\mathcal{V}_e^5, \mathcal{M}_e^5)$ and $(\mathcal{V}_e^6, \mathcal{M}_e^6)$ due to (11). Moreover, if $(0, \mathcal{M}_e^2)$ does not satisfy (14), then it will be an unstable saddle point (one positive and one negative eigenvalue), hence, low resistance to disturbances.

We will now discuss the stability of $(\mathcal{V}_e^5, \mathcal{M}_e^5)$ and $(\mathcal{V}_e^6, \mathcal{M}_e^6)$, satisfying the condition $\mathcal{M}_e^{5,6} > h_R U_o$. From (13) we get that such configurations are asymptotically stable if:

$$A + B < 0, \quad (A + B)^2 - 4(AB - K) \geq 0, \quad AB > K \tag{15}$$

or

$$A + B < 0, \quad (A + B)^2 - 4(AB - K) < 0. \tag{16}$$

In the first case, we have a stable node, whereas, in the second one, a stable focus. Instead, the condition

$$A + B < 0, \quad (A + B)^2 - 4(AB - K) > 0, \quad AB < K \tag{17}$$

furnishes unstable saddle points of equilibrium.

From the analysis performed on the equilibrium solutions of equations (2)–(3), it appears a variety of possible attractors toward which the ecological system, described by the model, will tend to move. These possible scenarios depend strongly from the values of the environmental parameters and from their relations, so that we can obtain an insight on the evolutionary trend of real ecological systems.

4. Numerical Applications

In this section we propose some numerical applications based on data found from GIS, relatively to the district of Cremona (Italy). In particular, it has been chosen a territory defined by a limited density of edified areas (13%), by a good percentage of areas with a medium-high biopotentiality (59.5 %), by a medium-low connectivity and by not critical values for the parameters h e h_R, linked, respectively, to the permeability of natural and anthropic

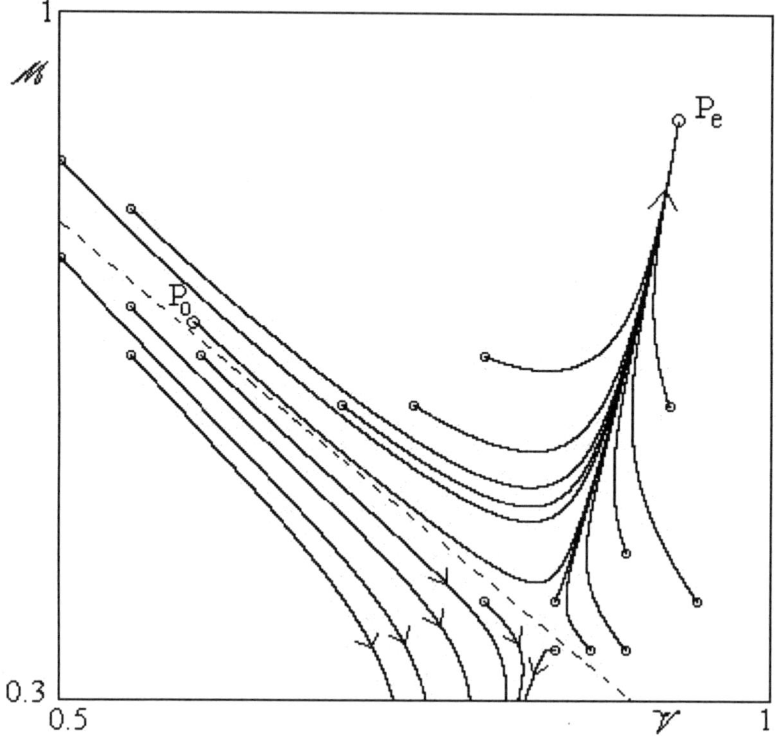

Fig. 2.   Basin of attraction of $P_e = (\mathcal{V}_e^6, \mathcal{M}_e^6)$ for different values of initial data

barriers, and to the environmental pressure of the low biopotentiality areas. Hence, we have considered the following numerical values in our simulations:

$$h_R = 0.44, \qquad U_o = 0.13, \qquad c = 0.454, \qquad h = 0.676.$$

These values correspond to the three equilibrium solutions of (9) with $H < 4/27$. We will examine only the case corresponding to a stable node, namely $(\mathcal{V}_e^6 = 0.936, \mathcal{M}_e^6 = 0.893)$ that represents a situation with larger metastabilty (larger resistance to disturbances) compared to the unstable cases.

Let us observe that, due to the presence of several equilibrium solutions, the trend toward each of them depends heavily from the initial data. From territorial data and the corresponding graph, we can fix the following initial conditions $(\mathcal{V}_0 = 0.595, \mathcal{M}_0 = 0.685) = P_o$ for the state variables of our model, corresponding to the selected land out of Cremona.

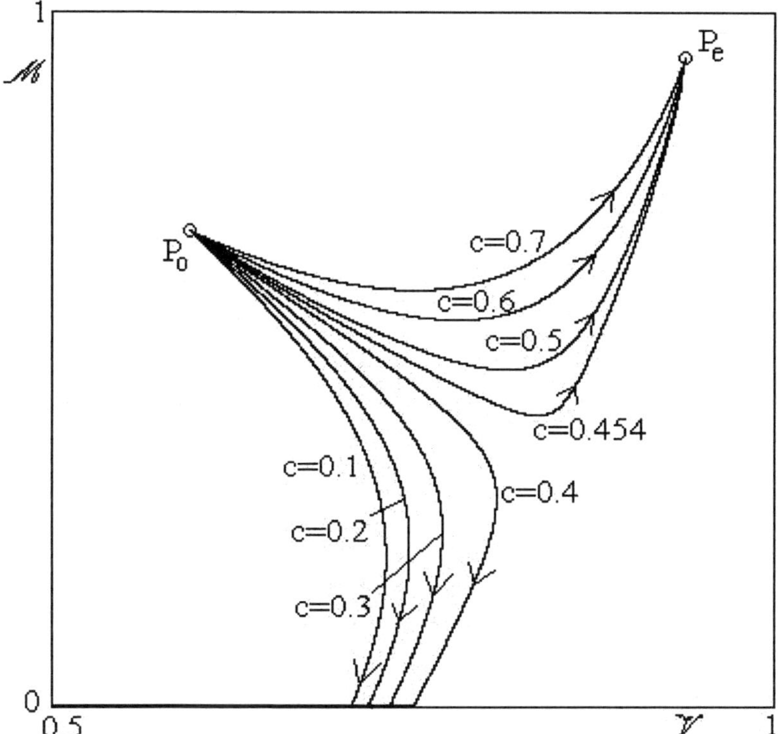

Fig. 3. Basin of attraction of $P_e = (\mathcal{V}_e^6, \mathcal{M}_e^6)$ for different values of c

In Fig.2 it is shown, in the phase plane $\big(\mathcal{V}(t), \mathcal{M}(t)\big)$, the basin of attraction of $P_e = (\mathcal{V}_e^6, \mathcal{M}_e^6)$: starting from the initial point P_o and from the others, on the right of the dashed line, the system evolves toward the end point of stable equilibrium P_e.

On the contrary, on the left of the dashed line one can find the set of initial values corresponding to the trend of the system toward unstable states.

In the next Fig.3, still in the phase plane $\big(\mathcal{V}(t), \mathcal{M}(t)\big)$ and starting from the same initial position P_o, it is shown, instead, the strong dependence of the behavior of the model from the connectivity parameter c.

As it can be understood, the proposed model may provide a significant quantitative tool in order to drive the planner decisions, which can be very often considered rather subjective and arbitrary. Thus, by trying to address the territorial settlement in the direction of maximizing the connectivity c and increasing the biodiversity (contained in the definition of the bioenergy

$\mathcal{M}$), the flux of bioenergy and the level of metastability will be incremented, driving, hence, the territory toward a sustainable environmental development.[4,6]

References

1. M. Abramowitz, I.A. Stegun, *Handbook of Mathematical Functions*, Dover, New York 1980.
2. A.L. Barabási, *La Scienza delle Reti*, Einaudi, Torino 2004.
3. M. Braun, *Differential Equations and Their Applications: An Introduction to Applied Mathematics*, Springer-Verlag, New York 1993.
4. P. Fabbri, *Paesaggio, Pianificazione, Sostenibilità*, Alinea Editrice, Firenze 2003.
5. R.F. Forman, M. Godron, *Landscape Ecology*, Wiley & Sons, N.Y. 1986
6. V. Ingegnoli, *Fondamenti di Ecologia del Paesaggio*, CittaStudi, Milano 1993.
7. V. Ingegnoli, R.F. Forman, *Landscape Ecology: A widening Foundation*, Springer-Verlag, New York 2002.
8. R.V. O'Neill, A.R. Johnson, A.W. King, Landscape Ecology, vol. 3, n.3/4, 1989
9. R. Monaco, G. Servente, *Introduzione ai Modelli Matematici nelle Scienze Territoriali*, CELID, Torino 2006.

Acknowledgements

The paper is partially supported by INDAM-GNFM and the National Research Project COFIN 2003 "Non linear mathematical problems of wave propagation and stability in models of continuous media" (Prof. T. Ruggeri), and Minho University Mathematics Centre (Financiamento plurianual, CMAT-FCT).

ANALYSIS OF THE LORENZ SYSTEM
AND THE BÉNARD PROBLEM WITH ROTATION
VIA THE CANONICAL REDUCTION METHOD

S. LOMBARDO*, G. MULONE° and M. TROVATO[†]

*Dipartimento di Matematica e Informatica, Università di Catania,
Catania, 95125, Italy*
*E-mail: * lombardo@dmi.unict.it, °mulone@dmi.unict.it, † trovato@dmi.unict.it*

Dedicated to Prof. Antonio Greco on the occasion of his 65th birthday

The canonical reduction method,[1–3] is applied to studying the stability of the
zero solution of the Lorenz system with and without rotation. Some stability
results of the motionless state of the rotating Bénard problem are also given.

Keywords: Canonical reduction method; Lorenz system; rotating Bénard problem; Lyapunov stability.

1. Introduction

In the study of the zero solution to linear and nonlinear ordinary differential
systems, the "canonical reduction method", which transforms a linearized
system in an uncoupled system or in a canonical (Jordan) form, plays a
fundamental role.[4,5]

Let us consider the ODE system

$$\dot{\mathbf{x}} = A\mathbf{x} + N(\mathbf{x}), \tag{1}$$

where $\mathbf{x} = (x_1, x_2, \ldots, x_n)^T \in \mathbb{R}^n$, A is a constant $n \times n$ matrix and N is a
nonlinear vector which vanishes for $\mathbf{x} = \mathbf{0}$. It is easy to see that if we study
the stability of the zero solution with respect to the *classical energy*

$$E_0 := \frac{1}{2}[x_1^2 + x_2^2 + \cdots + x_n^2],$$

in general we do not reach the stability results of linearized instability
(eigenvalues) method.

In order to study the stability of the zero solution and reach the stability results of the classical linearized method, we transform the system in an equivalent one with the *canonical reduction method*.

We introduce a *transformation* matrix Q (also called *modal* matrix), it is a non singular matrix of eigenvectors and generalized eigenvectors (in the case of a multiple eigenvalue with different geometric and algebraic multiplicity) of A. Q^{-1} is its inverse. In the case of simple eigenvalues, Q is given by an n by n array such that the jth column is the jth eigenvector corresponding to the jth eigenvalue. If the jth eigenvalue is complex, the jth column and the $(j+1)th$ column in the eigenvectors array are the real and imaginary parts corresponding to the jth eigenvalue, see.[4,5] Then we define the change of variables

$$\mathbf{x} = Q\mathbf{X},$$

$\mathbf{X} = (X_1, X_2, \ldots, X_n)^T$, we obtain the new system

$$\dot{\mathbf{X}} = Q^{-1}AQ\mathbf{X} + Q^{-1}N(Q\mathbf{X}), \qquad (2)$$

where B is a matrix similar to the matrix A. Now if we use the new *optimal Lyapunov function*

$$E := \frac{1}{2}[X_1^2 + X_2^2 + \cdots + X_n^2],$$

we reach,in the nonlinear case, the optimal stability results of linearized (eigenvalues) method.

By using the definition of the *principal eigenvalue* of an elliptic operator, this method can be applied to some PDEs systems which include reaction-diffusion systems and convection problems in fluid dynamics.[1,2] Here some results are given for the rotating Bénard problem.[6]

The plan of the paper is as follows: in Section 2, by using the canonical reduction method, we study the stability of the "zero solution" of the Lorenz system with (and without) rotation. In Section 3, we give some optimal stability results for the Bénard problem with rotation.

2. The Lorenz system with rotation

Let us consider the Lorenz model for the rotating Rayleigh - Bénard problem (see Bhattacharjee and McKane,[7] Lorenz[8]). The ordinary differential system is obtained by expanding in series the Oberbeck - Boussinesq equations for the rotating Bénard problem (see Section 3 and the papers[9,10]) and retaining the first term in the velocity perturbation field and the first two terms in the temperature perturbation. We have

$$\dot{x} = \sigma(-x + \tau g + y)$$
$$\dot{g} = -\sigma(\tau x + g)$$
$$\dot{y} = rx - y - xz \tag{3}$$
$$\dot{z} = -bz + xy$$

where the positive parameters σ, r, $b > 1$ and τ are the Prandtl number, the ratio of the Rayleigh number and the critical Rayleigh number for the onset of stationary convection, the aspect ratio and a normalized Taylor number.[7]

It can be proved[2] that the system admits a *maximal compact attractor*. In fact, we have

$$\limsup |\mathbf{x}(t)| \leq \frac{b(r + \sigma)}{2\sqrt{l(b-1)}},$$

where

$$|\mathbf{x}(t)| = \sqrt{x^2 + g^2 + y^2 + z^2}, \, l = \min(1, \sigma).$$

The equilibrium solutions of Eq. (3) are the zero solution $O = (0, 0, 0, 0)$ for any value of σ, r , b and τ. If $r > 1 + \tau^2$ there are also the equilibrium points

$$C_1 = (\overline{x}, -\tau\overline{x}, \overline{x}(1 + \tau^2), \frac{\overline{x}^2(1 + \tau^2)}{b}), \quad \text{with} \quad \overline{x} = \sqrt{\frac{b(r - 1 - \tau^2)}{1 + \tau^2}},$$

$$C_2 = (-\overline{x}, \tau\overline{x}, -\overline{x}(1 + \tau^2), \frac{\overline{x}^2(1 + \tau^2)}{b}).$$

Here we study the stability of the *zero solution* with the canonical reduction method. To this end, we observe that the eigenvalues of the associated matrix A are obtained by solving the equation

$$(b + \lambda)[(1 + \lambda)(\sigma + \lambda)^2 - (\sigma + \lambda)r\sigma + (1 + \lambda)\sigma^2\tau^2] = 0. \tag{4}$$

It can be seen that whenever $\sigma \geq 1$ and $r < 1 + \tau^2$ all the solutions of Eq. (4) have negative real parts, i.e. we have *linear and local nonlinear stability*.

For the sake of simplicity, here we consider analytically the case $\sigma = 1$, while, for $\sigma \neq 1$ we solve the problem by numerical methods.[2]

If $\sigma = 1$, the eigenvalues are

$$\lambda_1 = -b, \quad \lambda_{2,3} = -1 \pm \sqrt{r - \tau^2}, \quad \lambda_4 = -1.$$

If $r \leq \tau^2$ the eigenvalues $\lambda_{2,3}$ are complex conjugate or $\lambda_2 = \lambda_3 = 1$. In these cases the same procedure as that used in Example 1 can be applied. Here we restrict ourselves to *more dangerous* case the case $r > \tau^2$. We have instability whenever

$$-1 + \sqrt{r - \tau^2} > 0 \iff r > 1 + \tau^2.$$

Therefore, the *critical linear instability parameter* r_c is given by $1 + \tau^2$.

Because the matrix A is not symmetric, it can be seen that the classical energy (with the best coupling parameter r) $E_{0r} = \dfrac{1}{2}[r(x^2 + g^2) + (y^2 + z^2)]$ gives global stability *only* for $r < 1$ while the *stabilizing effect of the rotation is lost*.

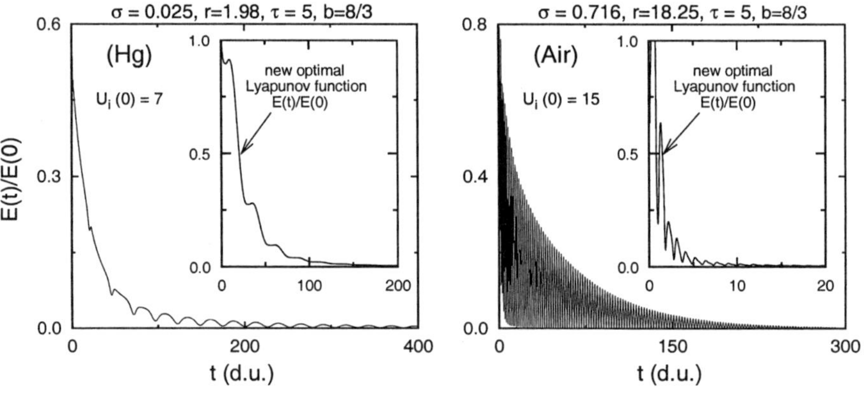

Fig. 1. Time evolution of the classical energy and the new optimal energy for the mercury, $\sigma = 0.025$, and the air, $\sigma = 0.716$, for given initial conditions and $b = 8/3$, $\tau = 5$. The values $r = 1.98$ (for the mercury) and $r = 18.25$ (for the air) near the critical instability parameters have been considered.

By using a transformation matrix Q and its inverse Q^{-1}, we easily obtain[2] the new system equivalent to Eq. (3):

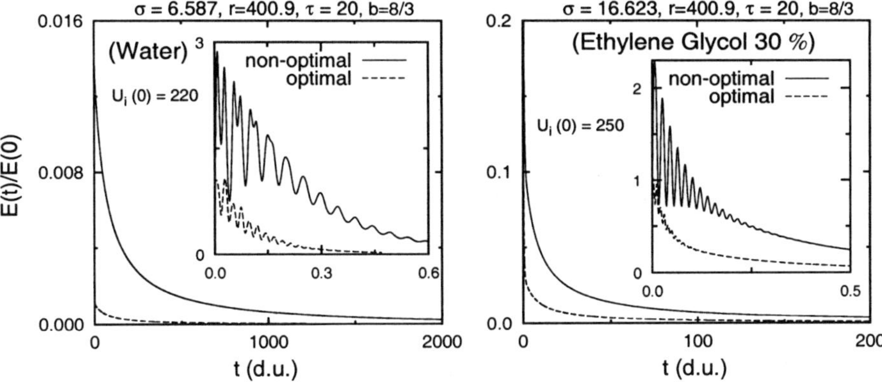

Fig. 2. Time evolution of the classical energy and the new optimal energy for the water, $\sigma = 6.587$, and the ethylene glycol 30%, $\sigma = 16.623$, for given initial conditions and $b = 8/3$, $\tau = 20$, near the critical value $r = 401$. The continuous curves correspond to classical (non-optimal) energy and the dashed curves correspond to optimal energy. In the zoom area the classical energy shows big oscillations (near the time t = 0).

$$\dot{X} = -bX + \frac{\sqrt{r - \tau^2}}{c_1 r}(c_2 G - c_3 Y)(c_2 G + c_3 Y + \tau c_4 Z)$$

$$\dot{G} = (-1 + \sqrt{r - \tau^2})G - \frac{c_1 X}{2c_2 \sqrt{r - \tau^2}}(c_2 G - c_3 Y)$$

$$\dot{Y} = (-1 - \sqrt{r - \tau^2})Y - \frac{c_1 X}{2c_3 \sqrt{r - \tau^2}}(c_2 G - c_3 Y) \tag{5}$$

$$\dot{Z} = -Z + \frac{\tau c_1 X}{c_4 r \sqrt{r - \tau^2}}(c_2 G - c_3 Y),$$

where c_i are not vanishing parameters to be chosen. By introducing the optimal Lyapunov function

$$E(t) = \frac{1}{2}\left(X^2 + G^2 + Y^2 + Z^2\right), \tag{6}$$

we have

$$\dot{E}(t) = -bX^2 + (-1 + \sqrt{r - \tau^2})G^2 - (1 + \sqrt{r - \tau^2})Y^2 - Z^2$$
$$+ (X\overline{n}_1 + G\overline{n}_2 + Y\overline{n}_3 + Z\overline{n}_4), \tag{7}$$

with $\overline{n}_i$ the nonlinear terms in the ith equation of the transformed system Eq. (5). We simplify the nonlinear terms by choosing

$$c_2 = c_3 = \frac{\sqrt{r}}{\sqrt{2(r - \tau^2)}}, \ c_4 = c_1 = 1.$$

We have

112

$$\dot{E}(t) \leq \beta E + kXZ(G - Y)\,, \tag{8}$$

where

$$\beta = 2\max(-1 + \sqrt{r - \tau^2}, -b),\ k = \frac{\tau(r + 1 - \tau^2)}{\sqrt{2r}(r - \tau^2)}\,.$$

From Eq. (8), by using the Cauchy inequality $XY \leq 1/2(X^2 + Y^2)$ we easily obtain

$$\dot{E}(t) \leq E[\beta + 2\sqrt{2}kE^{1/2}]\,. \tag{9}$$

If $\beta < 0$, *i.e.*,

$$\tau^2 < r < 1 + \tau^2$$

and

$$E(0) < \left(\frac{\beta}{2\sqrt{2}k}\right)^2\,, \tag{10}$$

by using a recursive argument, we have exponential decay (with a known radius of attraction (Eq. (10)))

$$E(t) \leq E(0)\exp\{[\beta + 2\sqrt{2}kE^{1/2}(0)]t\}\,. \tag{11}$$

Remark 2.1. *We note that, in the limit case $\tau = 0$, we have the classical Lorenz system, $k = 0$, and the global stability follows immediately from Eq. (9), (see also[5,8,11]).*

In general, to obtain (for arbitrary values of σ and r) the time evolution of the classical energy and the new optimal energy, it is necessary to solve numerically the non-linear ordinary differential equations Eq. (3) (for the variables $\mathbf{x} = (x, g, y, z)^T$) and the new system Eq. (5) (for the variables $\mathbf{X} = (X, G, Y, Z)^T$), by using a numerical procedure (like the Runge-Kutta method).

In particular, by fixing the values of b and τ, it is possible to calculate numerically[2] the spectrum of eigenvalues of matrix A as function of the parameters $\{r, \sigma\}$. We obtain that whenever $\sigma \geq 1$ the real parts of the eigenvalues are negative for $r < 1 + \tau^2$, while for $0 < \sigma < 1$ the critical instability parameter r_c depends on σ.

Thus following the general approach, also in the case of complex eigenvalues, we have considered different materials at temperature of 20C as, Mercury (Hg) ($\sigma = 0.025$), Air ($\sigma = 0.716$), Water ($\sigma = 6.587$) and the

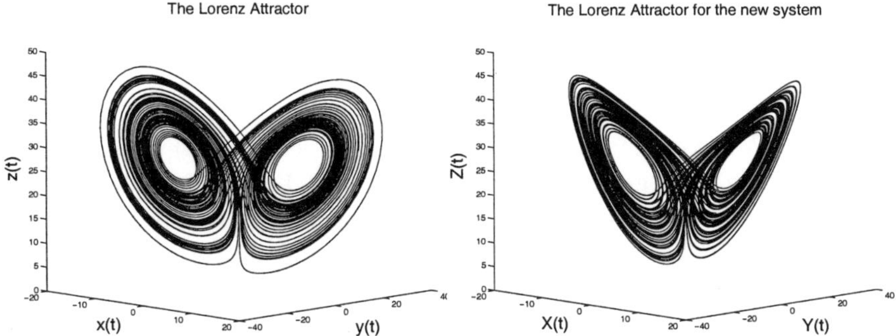

Fig. 3. Phase portraits of the *classical* Lorenz's chaotic attractor both for the usual system without rotation (on the left) and for the new system (on the right) obtained using the canonical reduction method. For the classical Lorenz system ($\tau = g = 0$), the usual values $\sigma = 10$, $b = 8/3$, $r = 28$, with the initial condition $x(0) = 15$, $y(0) = 20$, $z(0) = 30$, have been considered

Ethylene Glycol 30% ($\sigma = 16.623$). We have reported in Fig. 1-Fig. 2 the time evolution of the classical energy and the new optimal energy obtained near to the critical value r_c, for given initial conditions $U_i(0)$ and fixed values of b and τ.

Each curve is normalized to its initial value $E(0)$ to allow for a comparison of the different decay-time scales and the time is expressed in dimensionless units (d.u.)[7] A general property common to all the results is that the evolution of classical energy shows big oscillations that, in some cases (Hg, Air), extends to the whole range of temporal evolution (see Fig. 1 and the insets of Fig. 2). By contrast, the evolution of the new optimal energy shows a regular behavior (with some little oscillations) in the time and in many cases the shape of curves exhibits a decay-time scale faster than that of the classical energy. We have verified that these numerical results can be obtained for a large range of initial conditions $U_i(0)$.

In particular, an open problem is to prove the validity of *global stability* of the zero solution using different values of parameters σ, r with $\tau \neq 0$.

Our numerical results go in this direction, in fact in Fig. 1 and Fig. 2 the initial conditions outside of known radius of attraction for the system have been considered and in any case the evolution of energy exhibits a decay towards to the *zero* solution.

Finally, for the sake of completeness, in Fig. 3 we have reported the strange attractor both for the *classical Lorenz system* (without rotation) and for the *new system*, obtained via the canonical reduction method, when $\sigma = 10$, $b = 8/3$, $r = 28$ (and $\tau = 0$); the topological equivalence of the systems is

114

evident.

3. The Bénard problem with rotation

Let $d > 0$. We consider an infinite layer $\Omega_d = \mathbb{R}^2 \times (0, d)$ filled with an incompressible homogeneous newtonian fluid $\mathcal{F}$, heated from below, and subject to the action of a vertical gravity field $\mathbf{g} = -g\mathbf{k}$, where $\mathbf{k} = (0, 0, 1)$ is a unit vector. Let $\beta_1 > 0$ be the *adverse* temperature gradient which is maintained in the fluid. We also assume that the fluid is uniformly rotating about the vertical axis z with an angular velocity $\hat{\Omega}\mathbf{k}$, and denote by $Oxyz$ the cartesian frame of reference rotating about z with the same angular velocity. Let $\tilde{T}(x, y, 0) = \tilde{T}_0$, $\tilde{T}(x, y, d) = -\beta_1 d + \tilde{T}_0$ be the assigned temperatures on the planes $z = 0$ and $z = d$ respectively. The evolution equations of a nondimensional disturbance $(\mathbf{u}, \theta, p_1)$ of the basic conduction solution (see[9,10]) of the fluid in the rotating frame of reference are given by

$$\begin{cases} \mathbf{u}_t + \mathbf{u} \cdot \nabla\mathbf{u} = -\nabla p_1 + R\theta\mathbf{k} + T\mathbf{u} \times \mathbf{k} + \Delta\mathbf{u}, \quad \nabla \cdot \mathbf{u} = 0 \\ P_r(\theta_t + \mathbf{u} \cdot \nabla\theta) = Rw + \Delta\theta \end{cases} \tag{12}$$

in $\mathbb{R}^2 \times (0, 1) \times (0, \infty)$, where $\mathbf{u}$, θ, p_1 are the perturbations of the velocity, temperature and pressure fields, respectively (p_1 incorporates the centrifugal force); $\mathbf{u} = (u, v, w)$, R^2, T^2 and P_r are the Rayleigh, Taylor and Prandtl numbers, respectively. For the system (12) we adopt the initial and stress-free boundary conditions

$$\mathbf{u}(\mathbf{x}, 0) = \mathbf{u}_0(\mathbf{x}), \quad \theta(\mathbf{x}, 0) = \theta_0(\mathbf{x}), \qquad \mathbf{x} = (x, y, z) \in \mathbb{R}^2 \times (0, 1),$$

$$w = u_z = v_z = 0 = \theta = 0, \quad \text{on } z = 0, \, z = 1, \quad (x, y, t) \in \mathbb{R}^2 \times (0, \infty),$$

where $\theta_0(\mathbf{x})$, $\mathbf{u}_0(\mathbf{x})$ are prescribed regular fields with $\nabla \cdot \mathbf{u}_0 = 0$.
As usual, we assume that the perturbation fields are sufficiently smooth and are periodic functions in the x and y directions. We denote by

$$\Omega_p = (0, \frac{2\pi}{a_1}) \times (0, \frac{2\pi}{a_2}) \times (0, 1)$$

the periodicity cell, and by $a^2 = a_1^2 + a_2^2$ the wave number. Finally, in order to ensure uniqueness of the basic solution, we also require the "average velocity conditions"

$$\int_{\Omega_p} u \, d\Omega_p = \int_{\Omega_p} v \, d\Omega_p = 0.$$

It is well known (see for instance[9,12]) that both linear theory and the experiments show that rotation has a strong stabilizing effect (it inhibits the convection). The classical energy[10,13] fails to predict such an effect. Generalized energy functions have been introduced heuristically in the papers[14-18] and the stabilizing influence of rotation on convection has been proved.

Here we recall some results obtained in Mulone[6] to build an optimal Lyapunov function E_1 with the application of the canonical reduction method and the use of the *principal eigenvalue* of the Laplacian with zero boundary conditions. This Lyapunov function, as in the rotating Lorenz system, will give, via a variational problem, a critical parameter coincident with the critical linear instability parameter. The generalized energy of full nonlinear problem will be

$$V(t) = E_1(t) + bE_2(t),$$

where $b \geq 0$ and E_2 is a suitable complementary energy that must control the nonlinear terms.

If $P_r = 1$, by taking the curl and the double curl of the equation Eq. $(12)_1$ we obtain

$$\begin{cases} \Delta w_t = R\Delta^*\vartheta - T\zeta_z + \Delta\Delta w + N_1 \\ \zeta_t = Tw_z + \Delta\zeta + N_2 \\ \vartheta_t = Rw + \Delta\vartheta + N_3, \end{cases} \tag{13}$$

where $\zeta = \nabla \times \mathbf{u} \cdot \mathbf{k}$, $\Delta^* = \frac{\partial^2}{\partial x^2} + \frac{\partial^2}{\partial y^2}$ and N_i, $i = 1, 2, 3$, represent the nonlinear terms.

Now we take the partial derivative of the second equation with respect to z and obtain the new system

$$\begin{cases} \Delta w_t = R\Delta^*\vartheta - Th + \Delta\Delta w + N_1 \\ h_t = Tw_{zz} + \Delta h + N_4 \\ \vartheta_t = Rw + \Delta\vartheta + N_3, \end{cases} \tag{14}$$

where $h = \zeta_z$ and N_4 is the corresponding nonlinear term. For this system we have the boundary conditions $w = 0$, $\Delta w = 0$, $\vartheta = 0$, $h = 0$ on $z = 0$, $z = 1$.

The eigenvalues (in general complex numbers) of the associated linearized system are given by

$$\lambda_1 = -s, \quad \lambda_{\pm} = -s \pm \sqrt{\frac{R^2a^2 - T^2n^2\pi^2}{s}},$$

where $s = n^2\pi^2 + a^2$, $n \in \mathbb{N}^+$. We observe that if $R^2a^2 - T^2\pi^2 \leq 0$, the real part of the eigenvalues are negative and this implies linear stability.

116

We have instability if $R^2 a^2 - T^2 \pi^2 > 0$ and $\lambda_+ > 0$, *i.e.* (see[9])

$$R^2 > R_c^2 := \frac{T^2 \pi^2 + (\pi^2 + a^2)^3}{a^2} \tag{15}$$

with a^2 solution of the cubic equation

$$2(a^2)^3 + 3(a^2)^2 \pi^2 - \pi^2(\pi^4 + T^2) = 0.$$

R_c^2 is the critical Rayleigh number of linear instability.

By considering the linearized system associated to Eq. (14), because of the boundary conditions, and observing that all the even derivatives of h, w and ϑ vanish in $z = 0$, $z = 1$, we assume that the solutions are of this form

$$f(x, y, z, t) = f(x, y, t) \sin n\pi z,$$

with $f(x, y, t)$, of *plan form* type, *i.e.*, $\Delta^* f = -a^2 f$. By observing that the *principal eigenvalue* of the Laplacian is obtained for $n = 1$, we formally substitute Δf with $-(\pi^2 + a^2)f$, $\Delta^* f$ with $-a^2 f$, and $\frac{\partial^2}{\partial z^2} f$ with $-\pi^2 f$. We obtain the (ordinary) differential system

$$\begin{cases} w' = \dfrac{Ra^2}{s}\vartheta - sw + \dfrac{T}{s}h \\ h' = -sh - T\pi^2 w \\ \vartheta' = Rw - s\vartheta, \end{cases} \tag{16}$$

where $s = a^2 + \pi^2$. We assume that $R^2 a^2 - T^2 \pi^2 > 0$ (this is not a restriction, see[6]) and apply the canonical reduction method to the ODE system. Then we use the transformation of variables

$$\begin{cases} \phi = Rh + T\pi^2 \vartheta \\ \psi = \dfrac{\alpha}{2}w - \dfrac{T}{2\sqrt{s}}h - \dfrac{Ra^2}{2\sqrt{s}}\vartheta \\ \chi = \dfrac{\alpha}{2}w + \dfrac{T}{2\sqrt{s}}h + \dfrac{Ra^2}{2\sqrt{s}}\vartheta \end{cases} \tag{17}$$

where $\alpha = \sqrt{R^2 a^2 - T^2 \pi^2}$, to obtain the new equivalent system

$$\begin{cases} \phi' = -s\phi \\ \psi' = -(s + \dfrac{\alpha}{\sqrt{s}})\psi \\ \chi' = -(s - \dfrac{\alpha}{\sqrt{s}})\chi. \end{cases} \tag{18}$$

Now, recalling the expression of the Laplacian, and the other derivative operators we have used, with the inverse *formal transformation*

$$\Delta(\cdot) \longleftarrow -(\pi^2 + a^2)(\cdot), \quad \Delta^*(\cdot) \longleftarrow -a^2(\cdot), \quad \frac{\partial^2}{\partial z^2}(\cdot) \longleftarrow -\pi^2(\cdot),$$

we introduce the change of variables (for the linear and nonlinear PDE system)

$$\begin{cases} \phi = Rh - T\vartheta zz \\[2mm] \psi = -\dfrac{\alpha}{2s}\Delta w + \dfrac{1}{2\sqrt{s}}\Phi \\[2mm] \chi = -\dfrac{\alpha}{2s}\Delta w - \dfrac{1}{2\sqrt{s}}\Phi, \end{cases} \tag{19}$$

where $\Phi = -Th + R\Delta^*\vartheta$.

The optimal Lyapunov function we find is

$$E = \frac{\|\phi\|^2 + \|\psi\|^2 + \|\chi\|^2}{2}. \tag{20}$$

By noting that

$$\|\psi\|^2 + \|\chi\|^2 = \frac{\alpha^2}{2s^2}\|\Delta w\|^2 + \frac{1}{2s}\|\Phi\|^2, \tag{21}$$

we have

$$E = \frac{1}{2}[\|\phi\|^2 + \frac{\alpha^2}{2s^2}\|\Delta w\|^2 + \frac{1}{2s}\|\Phi\|^2]. \tag{22}$$

We note that this Lyapunov function has the same field variables as those introduced heuristically in other papers[6,14–18] Moreover we emphasize that here, with the reduction method, we obtain immediately the best Lyapounov coefficients $1, \dfrac{\alpha^2}{2s}, \dfrac{1}{2s}$. Thus, like in the rotating Lorenz system, we can obtain the coincidence between the critical linearized and the (conditional) nonlinear stability parameters.

Acknowledgments

Research partially supported by the University of Catania under a local contract, by the Italian Ministry for University and Scientific Research, PRIN: *"Problemi matematici non lineari di propagazione e stabilità nei modelli del continuo"*, and GNFM of INDAM.

References

1. S. Lombardo, G. Mulone and M. Trovato, *Rend. Circolo Mat. Palermo*, ser. II, Suppl. **78**, 173 (2006).
2. S. Lombardo, G. Mulone and M. Trovato, An operative method to define optimal Lyapunov functions in ODEs and in reaction-diffusion systems via the canonical reduction method (submitted).
3. G. Mulone and B. Straughan, *ZAMM* **86**, n. 7, 507 (2006).
4. M. W. Hirsch and S. Smale, *Differential equations, dynamical systems, and linear algebra*, (Academic Press, New York, 1974).
5. L. Perko, *Differential equations and dynamical systems*, 3rd edition (Springer-Verlag, New York, 2001).
6. G. Mulone, *Far East J. Appl. Math.* **15**, n.2, 117 (2004).
7. J.K. Bhattacharjee and A.J. Mckane, *J. Phys. A:Math. Gen.* **21**, L555 (1998).
8. E. N. Lorenz, *J. Atmos. Sci.* **20**, 130 (1963).
9. S. Chandrasekhar, *Hydrodynamic and hydromagnetic stability*, (Clarendon Press, Oxford, 1961).
10. D.D. Joseph, *Stability of fluid motions*, (Springer Tracts in Natural Philosophy, vols. 27 and 28. Springer, Berlin, 1976).
11. R. Temam, *Infinite-dimensional dynamical systems in mechanics and physics*, 2nd Edn. (Springer-Verlag, New York 1988).
12. E.L. Koschmieder, *Bénard cells and Taylor vortices*, (Cambridge Univ. Press, 1993).
13. J. Serrin, *Arch. Rational Mech. Anal.*, **3**, 1 (1959).
14. G.P. Galdi and B. Straughan, *Proc. Roy. Soc. Lond. A* , **402**, 257 (1985).
15. G. Mulone and S. Rionero, *J. Mat. Anal. App.* **144**, 109 (1989).
16. G. Mulone and S. Rionero, *Bull. Tech. Univ. Istanbul* **47**, 181 (1994).
17. G. Mulone and S. Rionero, *Continuum Mech. Termodyn.* **9**, 347 (1997).
18. R. Kaiser and L.X. Xu, *Nonlinear differ. equ. appl.* **5**, 283(1998).

LIE REMARKABLE PDEs

G. MANNO*, F. OLIVERI** and R. VITOLO*

*Department of Mathematics "E. De Giorgi", University of Lecce
via per Arnesano, 73100 Lecce, Italy
E-mail: gianni.manno@unile.it; raffaele.vitolo@unile.it

**Department of Mathematics, University of Messina
Salita Sperone 31, 98166 Messina, Italy
E-mail: oliveri@mat520.unime.it

Within the context of the inverse Lie problem the question whether there exist PDEs that are characterized by their Lie point symmetries may be addressed. In a recent paper the authors called these equations *Lie remarkable*. In this paper we exhibit various examples of Lie remarkable equations, including some multidimensional Monge-Ampère type equations.

Keywords: Lie symmetries of differential equations, jet spaces.

Dedicated to Antonio Greco
on the occasion of his 65th birthday.

1. Introduction

One of the most powerful tools for studying differential equations (DEs), either ordinary or partial, is provided by the theory of symmetries (see Refs. 1–9). Symmetries of DEs are (finite or infinitesimal) transformations of the independent and dependent variables and derivatives of the latter with respect to the former, with the further property of sending solutions into solutions. Among symmetries, there is a distinguished class, that of symmetries coming from a transformation of the independent and dependent variables, namely point symmetries.

Given a (system of) DE(s) the *direct* Lie problem consists in finding the admitted algebra of point symmetries. This task is accomplished by means of the Lie's algorithm, requiring the straightforward though tedious solution of an overdetermined system of DEs, possibly by using some computer algebra packages like Dimsym,[10] MathLie,[9] or Relie.[11]

The problem of finding the symmetries of a DE has associated a natural *inverse* problem, namely, the problem of finding the most general form of a DE admitting a given (abstract) Lie algebra as subalgebra of point symmetries. This problem was considered, for instance, in Refs. 12,13.

An aspect of this problem has been considered in Ref. 14, where the authors, starting from a given DE, found necessary and sufficient conditions for it to be uniquely determined by its point symmetries. By following the terminology already used in Refs. 14–17, we call such a DE *Lie remarkable*. A similar problem was also considered in Refs. 18,19.

In this paper we review some of the results obtained in Ref. 14, and give examples of Lie remarkable equations. Here we also treat the case of a multidimensional Monge-Ampère equation, *i.e.*, with more than two independent variables.

The plan of the paper is the following. In section 2, we introduce a DE of order r as a submanifold of a suitable jet space (of order r). Then we distinguish two types of Lie remarkable equations: *strongly* and *weakly* Lie remarkable equations. Strongly Lie remarkable equations are uniquely determined by their point symmetries; weakly Lie remarkable equations are equations which do not intersect other equations admitting the same symmetries. Then we report[14,16] necessary as well as sufficient conditions for an equation to be strongly or weakly Lie remarkable.

In section 3, we give various examples of either strongly or weakly Lie remarkable equations: they include equations of Monge-Ampère type, with two and three independent variables, and minimal surface equations.

2. Theoretical framework

Here we recall some basic facts regarding jet spaces (for more details, see Refs. 5,8,20) and the basic theory on DEs determined by their Lie point symmetries.[14]

All manifolds and maps are supposed to be C^∞. If E is a manifold then we denote by $\chi(E)$ the Lie algebra of vector fields on E. Also, for the sake of simplicity, all submanifolds of E are *embedded* submanifolds.

Let E be an $(n + m)$-dimensional smooth manifold and L an n-dimensional submanifold of E. Let (V, y^A) be a local chart on E. The coordinates (y^A) can be divided in two sets, $(y^A) = (x^\lambda, u^i)$, $\lambda = 1 \ldots n$ and $i = 1 \ldots m$, such that the submanifold L is locally described as the graph of a vector function $u^i = f^i(x^1, \ldots, x^n)$. In what follows, Greek indices run from 1 to n and Latin indices run from 1 to m unless otherwise specified.

The set of equivalence classes $[L]_p^r$ of submanifolds L having at $p \in E$

a contact of order r is said to be the *r-jet of n-dimensional submanifolds of E* (also known as extended bundles[5]), and is denoted by $J^r(E,n)$. If E is endowed with a bundle $\pi : E \to M$ where $\dim M = n$, then the r-th order jet $J^r\pi$ of local sections of π is an open dense subset of $J^r(E,n)$. We have the natural maps $j_r L \colon L \to J^r(E,n)$, $p \mapsto [L]_p^r$, and $\pi_{k,h} \colon J^k(E,n) \to J^h(E,n)$, $[L]_p^k \mapsto [L]_p^h$, $k \ge h$.

The set $J^r(E,n)$ is a smooth manifold whose dimension is

$$\dim J^r(E,n) = n + m \sum_{h=0}^{r} \binom{n+h-1}{n-1} = n + m \binom{n+r}{r}, \qquad (1)$$

whose charts are (x^λ, u_σ^i), where $u_\sigma^i \circ j_r L = \partial^{|\sigma|} f^i / \partial x^\sigma$, where $0 \le |\sigma| \le r$. On $J^r(E,n)$ there is a distribution, the contact distribution, which is generated by the vectors

$$D_\lambda \stackrel{\text{def}}{=} \frac{\partial}{\partial x^\lambda} + u_{\sigma\lambda}^j \frac{\partial}{\partial u_\sigma^j} \quad \text{and} \quad \frac{\partial}{\partial u_\tau^j},$$

where $0 \le |\sigma| \le r-1$, $|\tau| = r$ and $\sigma\lambda$ denotes the multi-index $(\sigma_1, \ldots, \sigma_{r-1}, \lambda)$. Any vector field $\Xi \in \chi(E)$ can be lifted to a vector field $\Xi^{(k)} \in \chi(J^k(E,n))$ which preserves the contact distribution. In coordinates, if $\Xi = \Xi^\lambda \partial/\partial x^\lambda + \Xi^i \partial/\partial u^i$ is a vector field on E, then its k-lift $\Xi^{(k)}$ has the coordinate expression

$$\Xi^{(k)} = \Xi^\lambda \frac{\partial}{\partial x^\lambda} + \Xi_\sigma^i \frac{\partial}{\partial u_\sigma^i}, \qquad (2)$$

where $\Xi_{\tau,\lambda}^j = D_\lambda(\Xi_\tau^j) - u_{\tau,\beta}^j D_\lambda(\Xi^\beta)$ with $|\tau| < k$.

A *differential equation $\mathcal{E}$ of order r on n-dimensional submanifolds of a manifold E* is a submanifold of $J^r(E,n)$. The manifold $J^r(E,n)$ is called the *trivial equation*. An *infinitesimal point symmetry* of $\mathcal{E}$ is a vector field of the type $\Xi^{(r)}$ which is tangent to $\mathcal{E}$.

Let $\mathcal{E}$ be locally described by $\{F^i = 0\}$, $i = 1 \ldots k$ with $k < \dim J^r(E,n)$. Then finding point symmetries amounts to solve the system

$$\Xi^{(r)}\left(F^i\right) = 0 \quad \text{whenever} \quad F^i = 0$$

for some $\Xi \in \chi(E)$.

We denote by $\mathrm{sym}(\mathcal{E})$ the Lie algebra of infinitesimal point symmetries of the equation $\mathcal{E}$.

By an *r-th order differential invariant* of a Lie subalgebra $\mathfrak{s}$ of $\chi(E)$ we mean a smooth function $I \colon J^r(E,n) \to \mathbb{R}$ such that for all $\Xi \in \mathfrak{s}$ we have $\Xi^{(r)}(I) = 0$.

122

The problem of determining the Lie algebra $\mathrm{sym}(\mathcal{E})$ is said to be the *direct Lie problem*. Conversely, given a Lie subalgebra $\mathfrak{s} \subset \chi(E)$, we consider the *inverse Lie problem*, *i.e.*, the problem of characterizing the equations $\mathcal{E} \subset J^r(E,n)$ such that $\mathfrak{s} \subseteq \mathrm{sym}(\mathcal{E})$.[1,21]

Definition 2.1. Let E be a manifold, $\dim E = n+m$, and let $r \in \mathbb{N}$, $r > 0$. An l-dimensional equation $\mathcal{E} \subset J^r(E,n)$ is said to be

(1) *weakly Lie remarkable* if $\mathcal{E}$ is the only maximal (with respect to the inclusion) l-dimensional equation in $J^r(E,n)$ passing at any $\theta \in \mathcal{E}$ admitting $\mathrm{sym}(\mathcal{E})$ as subalgebra of the algebra of its infinitesimal point symmetries;
(2) *strongly Lie remarkable* if $\mathcal{E}$ is the only maximal (with respect to the inclusion) l-dimensional equation in $J^r(E,n)$ admitting $\mathrm{sym}(\mathcal{E})$ as subalgebra of the algebra of its infinitesimal point symmetries.

Of course, a strongly Lie remarkable equation is also weakly Lie remarkable. Some direct consequences of our definitions are due. For each $\theta \in J^r(E,n)$ denote by $S_\theta(\mathcal{E}) \subset T_\theta J^r(E,n)$ the subspace generated by the values of infinitesimal point symmetries of $\mathcal{E}$ at θ. Let us set $S(\mathcal{E}) \overset{\mathrm{def}}{=} \bigcup_{\theta \in J^r(E,n)} S_\theta(\mathcal{E})$. In general, $\dim S_\theta(\mathcal{E})$ may change with $\theta \in J^r(E,n)$. The following inequality holds:

$$\dim \mathrm{sym}(\mathcal{E}) \geq \dim S_\theta(\mathcal{E}), \quad \forall \theta \in J^r(E,n), \tag{3}$$

where $\dim \mathrm{sym}(\mathcal{E})$ is the dimension, as real vector space, of the Lie algebra of infinitesimal point symmetries $\mathrm{sym}(\mathcal{E})$ of $\mathcal{E}$. If the rank of $S(\mathcal{E})$ at each $\theta \in J^r(E,n)$ is the same, then $S(\mathcal{E})$ is an involutive (smooth) distribution. A submanifold N of $J^r(E,n)$ is an *integral submanifold* of $S(\mathcal{E})$ if $T_\theta N = S_\theta(\mathcal{E})$ for each $\theta \in N$. Of course, an integral submanifold of $S(\mathcal{E})$ is an equation in $J^r(E,n)$ which admits all elements in $\mathrm{sym}(\mathcal{E})$ as infinitesimal point symmetries. The points of $J^r(E,n)$ of maximal rank of $S(\mathcal{E})$ form an open set of $J^r(E,n)$.[14] It follows that $\mathcal{E}$ can not coincide with the set of points of maximal rank of $S(\mathcal{E})$. The following theorems[14] can be proved.

Theorem 2.1.

(1) A necessary condition for the differential equation $\mathcal{E}$ to be strongly Lie remarkable is that

$$\dim \mathrm{sym}(\mathcal{E}) > \dim \mathcal{E}.$$

(2) A necessary condition for the differential equation $\mathcal{E}$ to be weakly Lie remarkable is that

$$\dim \operatorname{sym}(\mathcal{E}) \geq \dim \mathcal{E}.$$

In Ref. 14 also sufficient conditions have been established, that reveal useful when computing examples and applications.

Theorem 2.2.

(1) If $S(\mathcal{E})|_{\mathcal{E}}$ is an l-dimensional distribution on $\mathcal{E} \subset J^r(E,n)$, then $\mathcal{E}$ is a weakly Lie remarkable equation.

(2) Let $S(\mathcal{E})$ be such that for any $\theta \notin \mathcal{E}$ we have $\dim S_\theta(\mathcal{E}) > l$. Then $\mathcal{E}$ is a strongly Lie remarkable equation.

The next theorem[14] gives the relationship between Lie remarkability and differential invariants.

Theorem 2.3. *Let $\mathfrak{s}$ be a Lie subalgebra of $\chi(J^r(E,n))$. Let us suppose that the r-prolongation subalgebra of $\mathfrak{s}$ acts regularly on $J^r(E,n)$ and that the set of r-th order functionally independent differential invariants of $\mathfrak{s}$ reduces to a unique element $I \in C^\infty(J^r(E,n))$. Then the submanifold of $J^r(E,n)$ described by $\Delta(I) = 0$ (in particular $I = k$ for any $k \in \mathbb{R}$), with Δ an arbitrary smooth function, is a weakly Lie remarkable equation.*

To prove that a PDE is strongly or weakly Lie remarkable the following steps are required:

(1) determine its Lie point symmetries;
(2) determine the rank k of the distribution generated by its r–order prolongations and compare it with the dimension of the equation;
(3) determine the submanifolds where the rank of the distribution decreases.

3. Examples

In what follows we give some examples of Monge-Ampère equations (of various order) and minimal surface equations which are Lie remarkable. Since we deal with (infinitesimal) point symmetries of these equations, as we are interested to local aspects, we will interpret them as submanifolds of jets of a trivial bundle $\mathbb{R}^n \times \mathbb{R} \to \mathbb{R}^n$, with $n = 2,3$, which we denote, following our notation, by $J^r(\mathbb{R}^{n+1},n)$.

3.1. *Second order Monge-Ampère equation*

The 2nd order Monge-Ampère equation in 2 independent variables has been introduced by Ampère in 1815; in 1968, Boillat[22] discovered that it is the only second order equation being completely exceptional in the sense of Lax. The requirement of complete exceptionality has been used to derive Monge-Ampère equations involving more than 2 independent variables.[23–25]

Proposition 3.1 (Boillat, 1991). *Given an unknown field*

$$u(x_0, x_1, \ldots, x_n), \qquad (x_0 \text{ denoting the time}),$$

and its associated Hessian matrix $H = \left\| \frac{\partial^2 u}{\partial x_\alpha \partial x_\beta} \right\|$, *the most general 2nd order PDE being completely exceptional (and called Monge-Ampère equation) is provided by a linear combination of all minors extracted from H, with coefficients depending at most on x_α, u and first order derivatives of u.*

The classical Monge-Ampère equation written in the form

$$\kappa_1(u_{tt}u_{xx} - u_{tx}^2) + \kappa_2 u_{tt} + \kappa_3 u_{tx} + \kappa_4 u_{xx} + \kappa_5 = 0,$$

where the coefficients κ_i $(\kappa_1 \neq 0)$ are constant, through the substitution

$$u \to u - \frac{\kappa_4}{2\kappa_1}x^2 + \frac{\kappa_3}{2\kappa_1}xt - \frac{\kappa_2}{2\kappa_1}t^2$$

is mapped to

$$u_{tt}u_{xx} - u_{tx}^2 = \kappa, \qquad \kappa = \frac{4\kappa_2\kappa_4 - 4\kappa_1\kappa_5 - \kappa_3^2}{4\kappa_1^2}.$$

If $\kappa = 0$ we have the homogeneous Monge-Ampère equation for the surface $u(t, x)$ with zero Gaussian curvature.

Theorem 3.1. *Equation*

$$u_{tt}u_{xx} - u_{tx}^2 = \kappa \tag{4}$$

is weakly Lie remarkable if $\kappa \neq 0$, whereas it is strongly Lie remarkable if $\kappa = 0$.

Proof. Equation (4) is a hypersurface of $J^2(\mathbb{R}^3, 2)$. If $\kappa \neq 0$, equation (4) admits a 9-parameter group of point symmetries whose Lie algebra is spanned by the vector fields

$$\begin{aligned}
\Xi_1 &= \tfrac{\partial}{\partial x}, & \Xi_2 &= \tfrac{\partial}{\partial t}, & \Xi_3 &= \tfrac{\partial}{\partial u}, \\
\Xi_4 &= x\tfrac{\partial}{\partial t}, & \Xi_5 &= t\tfrac{\partial}{\partial x}, & \Xi_6 &= x\tfrac{\partial}{\partial u}, \\
\Xi_7 &= t\tfrac{\partial}{\partial u}, & \Xi_8 &= x\tfrac{\partial}{\partial x} + u\tfrac{\partial}{\partial u}, & \Xi_9 &= t\tfrac{\partial}{\partial t} + u\tfrac{\partial}{\partial u}.
\end{aligned}$$

The 2nd order prolonged vector fields give rise to a distribution of rank 7 (equation (4) is a 7–dimensional submanifold!) on the whole jet space provided we exclude the 5–dimensional submanifolds locally described by $u_{xx} = u_{tx} = u_{tt} = 0$ where the rank reduces to 5. Thus, non-homogeneous equation (4) is weakly Lie remarkable.

On the contrary, if $\kappa = 0$, equation (4) admits a 15–dimensional Lie algebra of point symmetries spanned by

$$\frac{\partial}{\partial a}, \quad a\frac{\partial}{\partial b}, \quad a\frac{\partial}{\partial b}, \quad a\left(\frac{\partial}{\partial x} + \frac{\partial}{\partial t} + \frac{\partial}{\partial u}\right), \quad \forall a, b \in \{x, t, u\}.$$

In this case, the 2nd order prolonged vector fields give rise to a distribution of rank 8 on $J^2(\mathbb{R}^3, 2)$ (which has dimension 8), provided we exclude the submanifold characterized by the equation itself (where the rank is at most 7; in fact on the submanifold $u_{xx} = u_{tx} = u_{tt} = 0$ the rank reduces to 5). Hence, the homogeneous equation for a surface with vanishing Gaussian curvature is strongly Lie remarkable. $\qquad\square$

3.2. *Monge-Ampère equation in 3 independent variables*

Consider the 2nd order Monge-Ampère equation in 3 independent variables:[23]

$$\kappa_1\left[u_{tt}(u_{xx}u_{yy} - u_{xy}^2) + u_{tx}(u_{ty}u_{xy} - u_{tx}u_{yy} + u_{ty}(u_{tx}u_{xy} - u_{ty}u_{xx}))\right]$$
$$+\kappa_2(u_{xx}u_{yy} - u_{xy}^2) + \kappa_3(u_{ty}u_{xy} - u_{tx}u_{yy}) + \kappa_4(u_{tx}u_{xy} - u_{ty}u_{xx})$$
$$+\kappa_5(u_{tt}u_{yy} - u_{ty}^2) + \kappa_6(u_{tx}u_{ty} - u_{tt}u_{xy}) + \kappa_7(u_{tt}u_{xx} - u_{tx}^2)$$
$$+\kappa_8 u_{tt} + \kappa_9 u_{tx} + \kappa_{10} u_{ty} + \kappa_{11} u_{xx} + \kappa_{12} u_{xy} + \kappa_{13} u_{yy} + \kappa_{14} = 0,$$

where κ_i $(i = 1, \ldots, 14)$ are taken constant.

The explicit determination of the infinitesimal generators of the admitted Lie group results quite complicated and the use of Computer Algebra packages reveals extremely memory consuming since the expression of the infinitesimals involves thousands of terms. Nevertheless, without loss of generality, it is possible to introduce the substitution

$$u \to u + \alpha_1 t^2 + \alpha_2 tx + \alpha_3 ty + \alpha_4 x^2 + \alpha_5 xy + \alpha_6 y^2,$$

α_i being suitable constants, so reducing the equation to the equivalent form

$$\kappa_1\left[u_{tt}(u_{xx}u_{yy} - u_{xy}^2) + u_{tx}(u_{ty}u_{xy} - u_{tx}u_{yy} + u_{ty}(u_{tx}u_{xy} - u_{ty}u_{xx}))\right]$$
$$+\kappa_2(u_{xx}u_{yy} - u_{xy}^2) + \kappa_3(u_{ty}u_{xy} - u_{tx}u_{yy}) + \kappa_4(u_{tx}u_{xy} - u_{ty}u_{xx})$$
$$+\kappa_5(u_{tt}u_{yy} - u_{ty}^2) + \kappa_6(u_{tx}u_{ty} - u_{tt}u_{xy}) + \kappa_7(u_{tt}u_{xx} - u_{tx}^2) = \kappa, \qquad (5)$$

In general this equation admits an 11–dimensional Lie algebra of point symmetries. Since the equation represents a 12–dimensional submanifold of $J^2(\mathbb{R}^4, 3)$ (which is a 13–dimensional manifold), it can not be neither strongly nor weakly Lie remarkable. However, the following theorem holds.

Theorem 3.2. *The 2nd order Monge-Ampère equation* (5), *when the coefficients are such that*

$$\kappa_1 = 1, \qquad \kappa_2\kappa_6^2 - \kappa_3\kappa_4\kappa_6 + \kappa_4^2\kappa_5 - (4\kappa_2\kappa_5 - \kappa_3^2)\kappa_7 = 0,$$

is weakly Lie remarkable.

Proof. In fact:

(1) the dimension of the submanifold described by (5) is 12;
(2) the Lie algebra of point symmetries is 13–dimensional;
(3) the 2nd order prolongations of the admitted vector fields give rise to a distribution of rank 12 provided that we exclude some submanifolds not contained in the equation itself.

Hence, due to theorem 2.2, equation (5) is weakly Lie remarkable. $\square$

Remark 3.1. It may be verified that the unique 2nd order differential invariant of the Lie symmetries of equation (5) is

$$I = (u_{tt}u_{xx}u_{yy} - u_{tt}u_{xy}^2 - u_{tx}^2 u_{yy} + 2u_{tx}u_{ty}u_{xy} - u_{ty}^2 u_{xx})$$
$$+\kappa_2(u_{xx}u_{yy} - u_{xy}^2) - \kappa_3(u_{tx}u_{yy} - u_{ty}u_{xy}) + \kappa_4(u_{tx}u_{xy} - u_{ty}u_{xx})$$
$$+\kappa_5(u_{tt}u_{yy} - u_{ty}^2) - \kappa_6(u_{tt}u_{xy} - u_{tx}u_{ty}) + \kappa_7(u_{tt}u_{xx} - u_{tx}^2),$$

whereupon if follows that they characterize the equation

$$\Delta(I) = 0 \quad \Rightarrow \quad I = \kappa, \quad \kappa \text{ constant}$$

More generally, some other 2nd order Monge-Ampère equations, involving more than 3 independent variables, are weakly Lie remarkable.

3.3. *Higher order Monge-Ampère equations*

The property of complete exceptionality has been used by Boillat[26] to determine higher order Monge-Ampère equations for the unknown $u(t, x)$.

By considering an equation of order $N > 2$, we need to consider the Hankel matrix

$$H = \begin{bmatrix} X_0 & X_1 & X_2 & \dots X_{M-1} & X_M \\ X_1 & X_2 & X_3 & \dots X_M & X_{M+1} \\ \dots & \dots & \dots & \dots\dots & \dots \\ X_{M-1} & X_M & X_{M+1} & \dots X_{2M-2} & X_{2M-1} \\ X_M & X_{M+1} & X_{M+2} & \dots X_{2M-1} & X_{2M} \end{bmatrix},$$

where $X_i = \dfrac{\partial^N u}{\partial t^i \partial x^{N-i}}$.

Theorem 3.3 (Boillat, 1992). *The most general nonlinear completely exceptional equation is given, if $N = 2M$, by a linear combination of all minors, including the determinant, of the Hankel matrix, whereas in the case where $N = 2M - 1$, we have to consider the linear combination of all minors extracted from the Hankel matrix where the last row has been removed.*

In both cases the coefficients of the linear combination are functions of t, x, u and its derivatives up to the order $N - 1$. Let us limit ourselves to the case where these coefficients are constant.

Consider the 3rd order Monge-Ampère equation

$$\widetilde{\kappa}_1(u_{ttx}u_{xxx} - u_{txx}^2) + \widetilde{\kappa}_2(u_{ttt}u_{xxx} - u_{ttx}u_{txx}) + \widetilde{\kappa}_3(u_{ttt}u_{txx} - u_{ttx}^2)$$
$$+\widetilde{\kappa}_4 u_{ttt} + \widetilde{\kappa}_5 u_{ttx} + \widetilde{\kappa}_6 u_{txx} + \widetilde{\kappa}_7 u_{xxx} + \widetilde{\kappa}_8 = 0.$$

The substitution

$$u \to u + \alpha_1 t^3 + \alpha_2 t^2 x + \alpha_3 t x^2 + \alpha_4 x^3$$

provides the equation

$$\kappa_1(u_{ttx}u_{xxx} - u_{txx}^2) + \kappa_2(u_{ttt}u_{xxx} - u_{ttx}u_{txx}) + \kappa_3(u_{ttt}u_{txx} - u_{ttx}^2) = \kappa. \quad (6)$$

Equation (6) describes an 11–dimensional submanifold of $J^3(\mathbb{R}^3, 2)$ (which is a 12–dimensional manifold); since the Lie algebra of its point symmetries is 10–dimensional, it can not be in general Lie remarkable.

Nevertheless, the following theorem may be proved.

Theorem 3.4. *The equation*

$$(u_{ttx}u_{xxx} - u_{txx}^2) + \lambda(u_{ttt}u_{xxx} - u_{ttx}u_{txx}) + \lambda^2(u_{ttt}u_{txx} - u_{ttx}^2) = \mu,$$

where

$$\lambda = \frac{\kappa_3}{\kappa_2}, \qquad \mu = \frac{\kappa \kappa_3}{\kappa_2^2},$$

obtained from (6) by choosing $\kappa_1 = \dfrac{\kappa_2^2}{\kappa_3}$, *is weakly Lie remarkable.*

Proof. In fact, the Lie algebra of point symmetries admitted is spanned by

$$\Xi_1 = \tfrac{\partial}{\partial t}, \qquad \Xi_2 = \tfrac{\partial}{\partial x}, \qquad \Xi_3 = \tfrac{\partial}{\partial u},$$
$$\Xi_4 = (2t - 3\lambda x)\tfrac{\partial}{\partial t} - x\tfrac{\partial}{\partial x}, \qquad \Xi_5 = \lambda^2 x\tfrac{\partial}{\partial t} + (2t - \lambda x)\tfrac{\partial}{\partial x},$$
$$\Xi_6 = \lambda x\tfrac{\partial}{\partial t} + x\tfrac{\partial}{\partial x} + 2u\tfrac{\partial}{\partial u}, \qquad \Xi_7 = t\tfrac{\partial}{\partial u}, \qquad \Xi_8 = x\tfrac{\partial}{\partial u},$$
$$\Xi_9 = t^2\tfrac{\partial}{\partial u}, \qquad \Xi_{10} = tx\tfrac{\partial}{\partial u}, \qquad \Xi_{11} = x^2\tfrac{\partial}{\partial u},$$
$$\Xi_{12} = F(t - \lambda x)\tfrac{\partial}{\partial u},$$

where F is an arbitrary function of $(t - \lambda x)$, and their 3rd order prolongations give rise, provided $F''' \neq 0$, to a distribution of rank 11, provided that we exclude some singular subsets. $\qquad\square$

Remark 3.2. It may be verified that the unique third order differential invariant is

$$I = (u_{ttx}u_{xxx} - u_{txx}^2) + \lambda(u_{ttt}u_{xxx} - u_{ttx}u_{txx}) + \lambda^2(u_{ttt}u_{txx} - u_{ttx}^2),$$

whereupon if follows that the operators characterize the equation

$$\Delta(I) = 0 \quad \Rightarrow \quad I = \mu.$$

Theorem 3.5. *The fourth order Monge-Ampère equation*

$$u_{tttt}(u_{ttxx}u_{xxxx} - u_{txxx}^2) + 2u_{tttx}u_{ttxx}u_{txxx} - u_{ttxx}^3 - u_{tttx}^2 u_{xxxx} = 0$$

is weakly Lie remarkable.

Proof.

(1) The previous equation describes a 16–dimensional submanifold of $J^4(\mathbb{R}^3, 2)$, which is a 17–dimensional manifold;
(2) The Lie algebra of point symmetries is 19–dimensional;
(3) The rank of 4th order prolongations give rise to a distribution of rank 16 provided we exclude a singular subset. $\qquad\square$

3.4. *Equation of minimal surface in $\mathbb{R}^{m+2}$*

Theorem 3.6. *The equation of minimal surface in $\mathbb{R}^{m+2}$*

$$(1 + |\mathbf{u}_y|^2)\mathbf{u}_{xx} - 2|\mathbf{u}_x| \cdot |\mathbf{u}_y|\mathbf{u}_{xy} + (1 + |\mathbf{u}_x|^2)\mathbf{u}_{yy} = \mathbf{0}, \quad \mathbf{u} \in \mathbb{R}^m, \qquad (7)$$

is nor strongly neither weakly Lie remarkable when $m = 2$ or $m = 3$, whereas is weakly Lie remarkable when $m = 1$ or $m = 4$.

Proof. If $m = 2$ or $m = 3$, the theorem follows immediately by dimensional reasons in view of theorem 2.1.

Then let us discuss the case $m = 1$.

In this case, equation (7) admits the following Lie algebra of point symmetries (which is formed by isometries and scaling):

$$\begin{aligned}
&\Xi_1 = \tfrac{\partial}{\partial x}, && \Xi_2 = \tfrac{\partial}{\partial y}, && \Xi_3 = \tfrac{\partial}{\partial u}, \\
&\Xi_4 = y\tfrac{\partial}{\partial x} - x\tfrac{\partial}{\partial y}, && \Xi_5 = u\tfrac{\partial}{\partial x} - x\tfrac{\partial}{\partial u}, && \Xi_6 = u\tfrac{\partial}{\partial y} - y\tfrac{\partial}{\partial u}, \\
&\Xi_7 = x\tfrac{\partial}{\partial x} + y\tfrac{\partial}{\partial y} + u\tfrac{\partial}{\partial u},
\end{aligned} \qquad (8)$$

The 2nd order prolongations of (8) give rise to a distribution of rank 7 on the equation, provided we exclude the differential equation of planes. Since equation (7) describes a 7–dimensional submanifold in $J^2(\mathbb{R}^3, 2)$ (which is an 8–dimensional manifold), from theorem 2.2 the result follows.

The case $m = 4$ is analogous to the case $m = 1$, then we omit computations. $\qquad\square$

Remark 3.3. The unique 2nd order differential invariant of (8) is:

$$I = \frac{\left((1 + u_y^2)u_{xx} - 2u_x u_y u_{xy} + (1 + u_x^2)u_{yy}\right)^2}{(1 + u_x^2 + u_y^2)(u_{xx}u_{yy} - u_{xy}^2)} = \frac{4H^2}{G}, \qquad (9)$$

where H is the scalar mean curvature and G the Gaussian curvature.

Then most general equation admitting the point symmetries of minimal surface equation in $\mathbb{R}^3$ is given by

$$\Delta(I) = 0, \quad \Rightarrow \quad I = \kappa, \quad \kappa \text{ constant.}$$

It is worth of noticing (see Ref. 27) that the equation

$$\frac{H^2}{G} = 1$$

is a strongly Lie remarkable equation characterized by the conformal algebra of $\mathbb{R}^3$.

Acknowledgments

Work supported by PRIN 2005/2007 ("Propagazione non lineare e stabilità nei processi termodinamici del continuo" and "Leggi di conservazione e termodinamica in meccanica dei continui e in teorie di campo"), GNFM, GNSAGA, Universities of Lecce and Messina.

References

1. G. W. Bluman, J. D. Cole, *Similarity methods of differential equations* (Springer, New York, 1974).
2. L. V. Ovsiannikov, *Group analysis of differential equations* (Academic Press, New York, 1982).
3. N. H. Ibragimov, *Transformation groups applied to mathematical physics* (D. Reidel Publishing Company, Dordrecht, 1985).
4. G. W. Bluman, S. Kumei, *Symmetries and differential equations* (Springer, New York, 1989).
5. P. J. Olver, *Applications of Lie Groups to Differential Equations*, 2nd edition (Springer, New York, 1991).
6. P. J. Olver, *Equivalence, Invariants, and Symmetry* (Cambridge University Press, New York, 1995).
7. N. H. Ibragimov, *Handbook of Lie group analysis of differential equations*, 3 volumes (CRC Press, Boca Raton, 1994, 1995, 1996).
8. A. V. Bocharov, V. N. Chetverikov, S. V. Duzhin, N. G. Khor'kova, I. S. Krasil'shchik, A. V. Samokhin, Yu. N. Torkhov, A. M. Verbovetsky and A. M. Vinogradov, *Symmetries and Conservation Laws for Differential Equations of Mathematical Physics*, I. S. Krasil'shchik and A. M. Vinogradov eds., Translations of Math. Monographs, **182**, Amer. Math. Soc. (1999).
9. G. Baumann, *Symmetry analysis of differential equations with Mathematica* (Springer, Berlin, 2000).
10. J. Sherring, Dimsym: symmetry determination and linear differential equation package, 1996 (available at the URL `http://www.latrobe.edu.au/mathstats/maths/department/dimsym/`).
11. F. Oliveri, ReLie: A Reduce program for Calculating Lie Point Symmetries of Differential Equations, 2001 (available at the URL `http://mat520.unime.it/oliveri`).
12. W. I. Fushchych, I. Yehorchenko, Second-order differential invariants of the rotation group $O(n)$ and of its extension: $E(n)$, $P(1,n)$, $G(1,n)$, *Acta Appl. Math.*, **28**, 69–92 (1992).
13. G. Rideau, P. Winternitz, Nonlinear equation invariant under the Poincaré, similitude and conformal groups in two-dimensional space-time, *J. Math. Phys.*, **31**, 1095–1105 (1990).
14. G. Manno, F. Oliveri, R. Vitolo, On differential equations characterized by their Lie point symmetries, to appear in *J. Math. Anal. Appl.* (2006).
15. F. Oliveri, Lie symmetries of differential equations: direct and inverse problems, *Note di Matematica*, **23**, 195–216 (2004/2005).
16. G. Manno, F. Oliveri, R. Vitolo, On an inverse problem in group analysis of PDEs: Lie-remarkable equations, in *Wascom 2005, Proc. XIII Int. Conf. on Waves and Stability in Continuous Media*, (R. Monaco, G. Mulone, S. Rionero, T. Ruggeri editors), World Scientific, Singapore, 420–432 (2005).
17. F. Oliveri, Sur une propriété remarquable des équations de Monge-Ampère, *Suppl. Rend. Circ. Mat. Palermo "Non Linear Hyperbolic Fields and Waves – A tribute to Guy Boillat"*, Ser. II, N. 78, 243–257 (2006).
18. V. Rosenhaus, The unique determination of the equation by its invariance

group and field-space symmetry, *Algebras, Groups and Geometries*, **3**, 148–166 (1986).

19. V. Rosenhaus, Groups of invariance and solutions of equations determined by them, *Algebras, Groups and Geometries*, **5**, 137–150 (1988).
20. D. J. Saunders, *The Geometry of Jet Bundles* (Cambridge University Press, 1989).
21. W. I. Fushchych, *Collected Works* (Kyiv, 2000).
22. G. Boillat, Le champ scalaire de Monge-Ampère, *Det. Kgl. Norske Vid. Selsk. Forth.*, **41**, 78–81 (1968).
23. T. Ruggeri, Su una naturale estensione a tre variabili dell'equazione di Monge-Ampère, *Rend. Acc. Naz. Lincei*, **55**, 445–449 (1973).
24. A. Donato, U. Ramgulam, C. Rogers, The (3+1)-dimensional Monge-Ampère equation in discontinuity wave theory: application of a reciprocal transformation, *Meccanica*, **27**, 257–262 (1992).
25. G. Boillat, Sur l'équation générale de Monge-Ampère à plusieurs variables, *C. R. Acad. Sci. Paris*, Sér. I. Math., **313**, 805–808 (1991).
26. G. Boillat, Sur l'équation générale de Monge–Ampère d'ordre supérieur, *C. R. Acad. Sci. Paris*, Sér. I. Math., **315**, 1211–1214 (1992).
27. G. Manno, F. Oliveri, R. Vitolo, Differential equations uniquely determined by algebras of point symmetries, Preprint (2006).

ON TWO-PULSE AND SHOCK EVOLUTION IN A CLASS OF IDEALLY HARD ELASTIC MATERIALS

A. MENTRELLI[1], C. ROGERS[2], T. RUGGERI[1], W. K. SCHIEF[3]

[1] *Research Centre of Applied Mathematics (CIRAM), University of Bologna, Italy*
[2] *School of Mathematics, the University of New South Wales, Sydney, Australia &*
Australian Research Council Centre of Excellence for Mathematics and Statistics of
Complex Systems
[3] *Institut für Mathematik, Technische Universität Berlin, Germany & Australian*
Research Council Centre of Excellence for Mathematics and Statistics of Complex
Systems

Pulse and shock interaction is investigated for a class of ideally hard elastic materials. The latter admit the soliton-like property that disturbances emerge asymptotically unaltered in shape and speed following the interaction process.

Keywords: Hard elastic materials; Solitonic behaviour; Nonlinear wave propagation.

1. Introduction

Loewner[1] introduced a class of Bäcklund transformations (BTs) which were applied to the hodograph equations of gasdynamics in order to obtain reduction to appropriate canonical forms in subsonic, trans-sonic and supersonic flow régimes. Such reduction was derived for certain multi-parameter model constitutive laws which were used to approximate real gas behaviour. Particular such laws obtained independently were applied by Dombrovskii[2] to solve a variety of boundary value problems. Analogous model constitutive laws may be constructed 'mutatis mutandis' in nonlinear elastostatics[3] and elastodynamics[4]. In the latter context, such model stress-strain laws were used by Varley *et al.*[5,6] to investigate the transmission and reflexion of pulses in bounded nonlinear elastic media. Subsequently, it was established in Ref. 7 that such materials have the remarkable property that pulses travelling in the opposite direction, despite distorting as they propagate, emerge unchanged from the interaction process. This is reminiscent of solitonic interaction of shapes of permanent form. Indeed, in recent work[8] a solitonic, connection has been made between the model stress-strain laws of Ref. 5,6 and the integrable sinh-Gordon equation. This connection was exploited to construct novel stress-strain laws involving elliptic functions which allow interior changes in concavity such as are encountered in superelastic materials like nitinol[9,10].

In Ref. 11, pulse propagation and shock evolution was investigated for a class of ideally hard, inhomogeneous elastic materials associated with pseudospherical surfaces. Here, two-pulse and shock interaction are considered for another class of ideally hard but homogeneous elastic materials for which the hodograph system may be reduced to the classical wave equation by a single BT. The constitutive laws involve a subclass of those introduced in Ref. 5, gasdynamic analogues of which were obtained earlier in the work of Loewner[1]. The nonlinear interaction of pulses which evolve into shocks is examined for a variety of initial data and the soliton-like interaction placed in evidence.

2. The Infinitesimal Bäcklund Transformations

Here, we consider the one-dimensional deformation $x = x(X, t)$ of a nonlinear elastic medium with stress-strain law

$$T = \Sigma(e), \tag{1}$$

where T is the stress and

$$e = x_X - 1 \tag{2}$$

is the strain. In the above, x and X are, in turn, Eulerian and Lagrangian coordinates and subscripts denote derivatives. The governing Lagrangian equations of motion consist of the compatibility condition

$$e_t = v_X, \tag{3}$$

where

$$v = x_t$$

is the velocity, together with the momentum equation

$$T_X = \rho_0 v_t. \tag{4}$$

Here, ρ_0 is the density of the medium in its undeformed state. Insertion of (1) into (4) yields

$$A^2 e_X = v_t, \tag{5}$$

134

where

$$A = \sqrt{\rho_0^{-1} \frac{d\Sigma}{de}}$$

is the signal speed (and $\pm A$ are the characteristic velocities of the hyperbolic system (3), (5)).

On introduction of the hodograph transformation, wherein X, t are taken as new dependent variables and v, e as new independent variables, it is seen that

$$v_X = Jt_e, \quad v_t = -JX_e$$
$$e_x = Jt_v, \quad e_t = JX_v, \tag{6}$$

where $J = J(v, e; X, t)$ is the Jacobian and it is required that

$$0 < |J| < \infty.$$

Use of the relations (6) in (3) and (5) produces the hodograph system

$$\begin{pmatrix} X \\ t \end{pmatrix}_\sigma = \begin{pmatrix} 0 & A \\ A^{-1} & 0 \end{pmatrix} \begin{pmatrix} X \\ t \end{pmatrix}_\tau, \tag{7}$$

where

$$\frac{d\sigma}{de} = A, \quad \tau = v. \tag{8}$$

Loewner[1], in the context of gasdynamics, introduced a class of matrix BTs

$$\boldsymbol{\Phi}'_\tau = \boldsymbol{\Lambda}_1 \boldsymbol{\Phi}_\tau + \mathbf{A}_1 \boldsymbol{\Phi} + \mathbf{B}_1 \boldsymbol{\Phi}',$$
$$\boldsymbol{\Phi}'_\sigma = \boldsymbol{\Lambda}_2 \boldsymbol{\Phi}_\sigma + \mathbf{A}_2 \boldsymbol{\Phi} + \mathbf{B}_2 \boldsymbol{\Phi}', \tag{9}$$

$$\sigma' = \sigma, \quad \tau' = \tau$$

which take a hodograph system

$$\boldsymbol{\Phi}_\sigma = \mathbf{S} \boldsymbol{\Phi}_\tau \tag{10}$$

to an associated system

$$\boldsymbol{\Phi}'_\sigma = \mathbf{S}' \boldsymbol{\Phi}'_\tau. \tag{11}$$

The requirement that $\boldsymbol{\Phi}'$ be a solution of (11) yields

$$\mathbf{S}' \boldsymbol{\Lambda}_1 = \boldsymbol{\Lambda}_2 \mathbf{S}, \tag{12}$$
$$\mathbf{A}_2 = \mathbf{S}' \mathbf{A}_1, \tag{13}$$
$$\mathbf{B}_2 = \mathbf{S}' \mathbf{B}_1, \tag{14}$$

while the integrability condition applied to (9) produces the relations

$$\Lambda_{1\sigma} + \mathbf{B}_1\Lambda_1\mathbf{S} + \mathbf{A}_1\mathbf{S} = \Lambda_{1\tau}\mathbf{S} + \mathbf{B}_2\Lambda_1 + \mathbf{A}_2, \tag{15}$$

$$\mathbf{A}_{1\sigma} - \mathbf{A}_{2\tau} + \mathbf{B}_1\mathbf{A}_2 - \mathbf{B}_2\mathbf{A}_1 = \mathbf{0}, \tag{16}$$

$$\mathbf{B}_{1\sigma} - \mathbf{B}_{2\tau} + \mathbf{B}_1\mathbf{B}_2 - \mathbf{B}_2\mathbf{B}_1 = \mathbf{0}, \tag{17}$$

together with $\Lambda_2 = \Lambda_1$. Thus, any solution of the nonlinear system (12)–(17) determines a transformation between the linear systems (10) and (11). It turns out that the system (12)–(17) may be entirely characterised in terms of what, in a solitonic context, are eigenfunctions and adjoint eigenfunctions[12].

Here, we are concerned with the action of a particular single BT of the type (9), namely

$$\begin{pmatrix} X' \\ t' \end{pmatrix}_\tau = \begin{pmatrix} a_{11} & 0 \\ 0 & a_{22} \end{pmatrix} \begin{pmatrix} X \\ t \end{pmatrix}_\tau + \begin{pmatrix} 0 & c_{12} \\ c_{21} & 0 \end{pmatrix} \begin{pmatrix} X \\ t \end{pmatrix},$$

$$\begin{pmatrix} X' \\ t' \end{pmatrix}_\sigma = \begin{pmatrix} a_{11} & 0 \\ 0 & a_{22} \end{pmatrix} \begin{pmatrix} X \\ t \end{pmatrix}_\sigma + \begin{pmatrix} c_{21} & 0 \\ 0 & c_{12} \end{pmatrix} \begin{pmatrix} X \\ t \end{pmatrix},$$

where

$$a_{11}a_{22} = \text{const} = a \neq 0,$$

which takes the hodograph system (7) to the canonical form

$$\begin{pmatrix} X' \\ t' \end{pmatrix}_\sigma = \begin{pmatrix} 0 & 1 \\ 1 & 0 \end{pmatrix} \begin{pmatrix} X' \\ t' \end{pmatrix}_\tau, \tag{18}$$

where

$$a_{11} = \sqrt{a/A} \tag{19}$$

is governed by the Riccati equation

$$a_{11\sigma} + \alpha a_{11}^2 + \beta = 0 \tag{20}$$

with $c_{12} = a\alpha$, $c_{21} = -\beta$. On insertion of (19) into (20), the signal speed relation

$$A_\sigma = \mu A^{1/2} + \nu A^{3/2} \tag{21}$$

as set down in Ref. 5 is retrieved. It is interesting to remark that the above relation may also be obtained via single '1-soliton-like' solutions of an integrable sinh-Gordon equation associated with the action of infinitesimal BTs[8]. Indeed, as shown in Ref. 12, the finite Loewner transformations (9) may be obtained by integration of infinitesimal BTs.

3. Ideally Hard Elastic Materials

If we set $\mu = 0$ in (21) then integration produces the signal speed relation

$$\frac{A}{A_0} = \left(1 - \frac{\nu\sigma}{2A_0^{-1/2}}\right)^{-2}, \tag{22}$$

where $A_0 = A\big|_{\sigma=0}$. The expressions (22) and (8) now give the stress-strain laws parametrically via

$$\frac{T}{\rho_0 A_0^2 e_1} = 3\left[\left(1 - \frac{\sigma}{3A_0 e_1}\right)^{-1} - 1\right], \tag{23}$$

$$\frac{e}{e_1} = 1 - \left(1 - \frac{\sigma}{3A_0 e_1}\right)^3, \tag{24}$$

where $e_1 = 2/(3A_0^{3/2}\nu)$ and $T|_{\sigma=0} = e|_{\sigma=0} = 0$. Elimination of σ between (23) and (24) produces the explicit stress-strain law

$$\frac{T}{\rho_0 A_0^2 e_1} = 3\left[\left(1 - \frac{e}{e_1}\right)^{-1/3} - 1\right]. \tag{25}$$

Here, we require that $\nu > 0$ so that $e_1 > 0$ and it is seen that $e \to e_1$ as $T \to \infty$, where e_1 is a locking strain. Thus, the constitutive law (25) models ideally hard elastic materials. It has been used in Ref. 5 to correlate with the dynamic response of clays and saturated soils. Here, our purpose is to investigate the nonlinear interaction of pulses in such materials corresponding to a variety of initial data.

4. Integration of the Hodograph System

It is known[12] that the finite Loewner transformation (9) may be inverted explicitly. In particular, the general solution of the hodograph system (7) with A given by (22), that is

$$A = A_0(1 - \alpha\sigma)^{-2}, \quad \alpha = \frac{1}{3e_1 A_0},$$

may be expressed in terms of the general solution of the reduced hodograph system (18). Since the latter is equivalent to the wave equation, we obtain a representation of the general solution of (7) in terms of two arbitrary functions of one variable. Indeed, it may be directly verified that

$$X = \frac{f'(\sigma + v) + g'(\sigma - v)}{1 - \alpha\sigma}$$

$$t = \frac{1 - \alpha\sigma}{A_0}[f'(\sigma + v) - g'(\sigma - v)] + \frac{\alpha}{A_0}[f(\sigma + v) - g(\sigma - v)].$$

The latter is the 'hyperbolic analogue' of the solution of the hodograph system given in Ref. 13.

If we impose the natural initial conditions (cf. Section 5)

$$e(X, t = 0) = e_0(X), \quad v(X, t = 0) = v_0 = 0 \tag{26}$$

then $(26)_2$ yields

$$0 = (1 - \alpha\rho)[f'(\rho) - g'(\rho)] + \alpha[f(\rho) - g(\rho)]$$

so that

$$f(\rho) = g(\rho) + c(1 - \alpha\rho).$$

The function g is then determined by the remaining initial condition $(26)_1$ with

$$e = e_1[1 - (1 - \alpha\sigma)^3],$$

that is, the relation

$$e_1[1 - (1 - \alpha\rho)^3] = e_0\left(\frac{2g'(\rho)}{1 - \alpha\rho} - \frac{c\alpha}{1 - \alpha\rho}\right).$$

At any station X for which $e_0'(X) \neq 0$, the local solution of the above relation is therefore given by

$$g(\rho) = \frac{1}{2}c\alpha\rho + \frac{1}{2}\int (1 - \alpha\rho)h(e_1[1 - (1 - \alpha\rho)^3])\,d\rho$$

with $h \circ e_0 = \mathrm{id}$.

5. Pulse Interaction

We now consider pulse interaction corresponding to various initial data in the class of ideally hard elastic materials governed by the nonlinear system (3), (5) with nonlinear stress-strain law (25).

The pulse interaction is studied with a numerical approach, making use of a previously developed general-purpose code useful for solving hyperbolic systems of nonlinear balance laws. The implemented algorithm is a

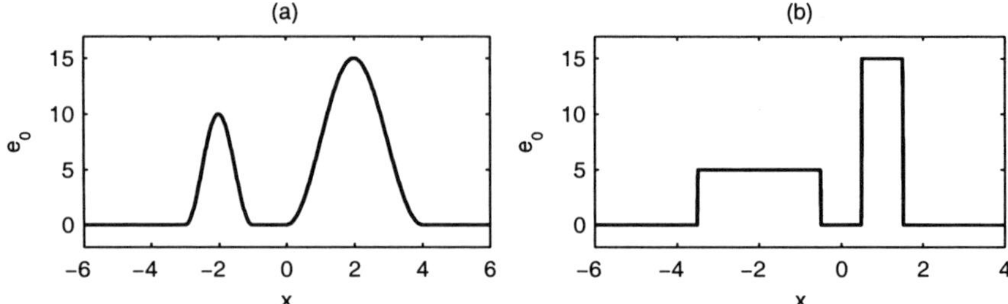

Fig. 1.   Initial data for the strain, e_0, with (a) two continuous sinusoidal pulses and (b) two rectangular pulses.

slightly modified version of the *Uniformly accurate Central Scheme of order 2* (UCS2), belonging to the central scheme family, developed by Liotta, Romano and Russo[14] and described elsewhere[15].

In the following, results are discussed for ideally hard elastic materials with two different values of the locking strain e_1. For the first of these two materials, characterised by a locking strain $e_1 = 20$, results on pulse interaction are presented for different kinds of initial pulses in the strain profile, namely continuous sinusoidal and discontinuous rectangular signals, respectively, as illustrated in Figure 1. For the second hard elastic material, described by a locking strain $e_1 = 60$, results are presented for sinusoidal initial data as shown in Figure 1(a). In all three cases, the initial profile of the velocity, v, is assumed to be $v_0 \equiv v(X, t = 0) = 0$. The calculations are based on the stress-strain law (25) with $A_0 = \rho_0 = 1$.

In order to appreciate the soliton-like behaviour of the pulses, the evolution of the strain and velocity profiles is compared with the superposition of the corresponding evolutions of the single pulses taken separately. Indeed, even though the interaction process results in a translation of the pulses relative to the position of the non-interacting pulses, it is seen that in all the three cases discussed here, after the interaction, the disturbances emerge asymptotically unaltered in shape and velocity: in Figures 2 and 4, where after the interaction process the emerging pulses exhibit a shock, as well as in Figure 3, where the shock formation has not yet occurred after the interaction. In the cases presented in Figures 2 and 4, it is also evident that the pulses emerging from the interaction process have the same shape as the non-interacting pulses.

It is worth noticing that, according to the theoretical results by Liu[16], the solution of a system of conservation laws subject to initial data with

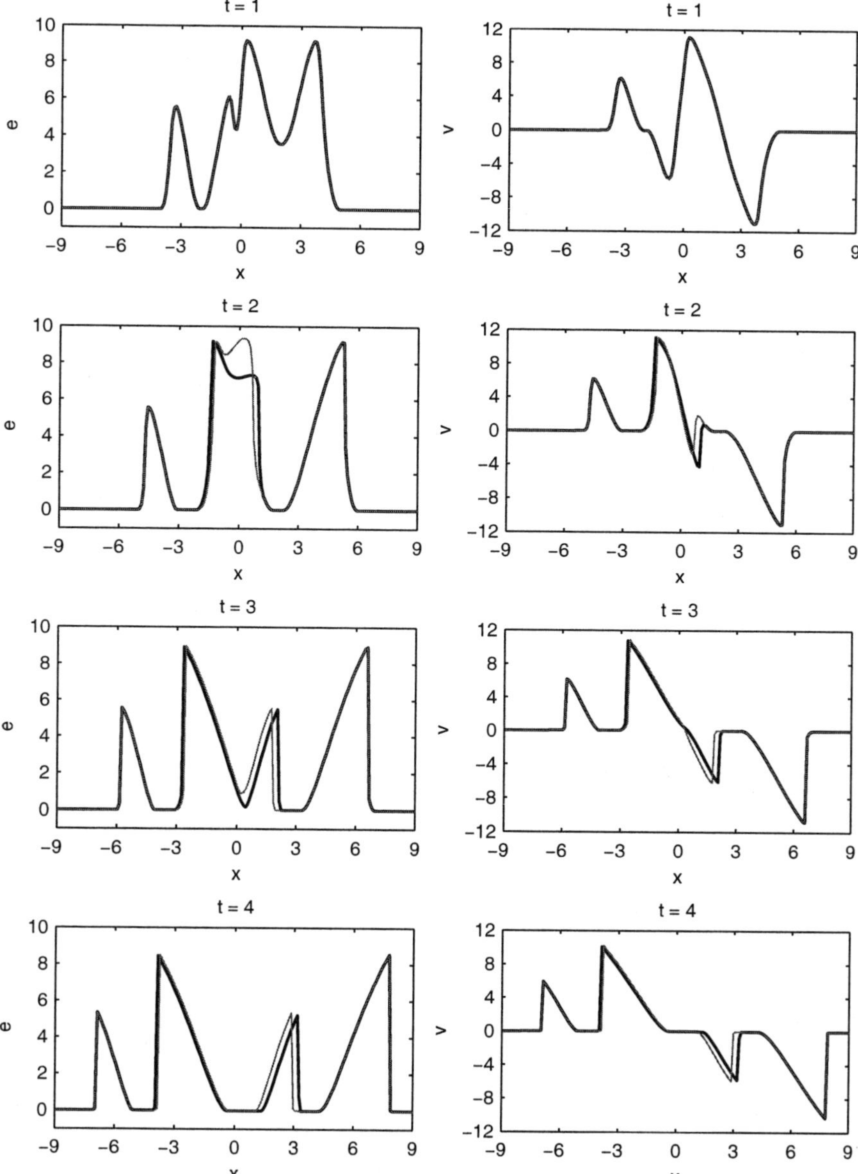

Fig. 2. Evolution of strain (left) and velocity (right) fields for initial data with two pulses (thick line), compared to the superposition of the evolutions for initial data with the two pulses taken separately (thin line), in the case $e_1 = 20$ under initial conditions shown in Figure 1(a).

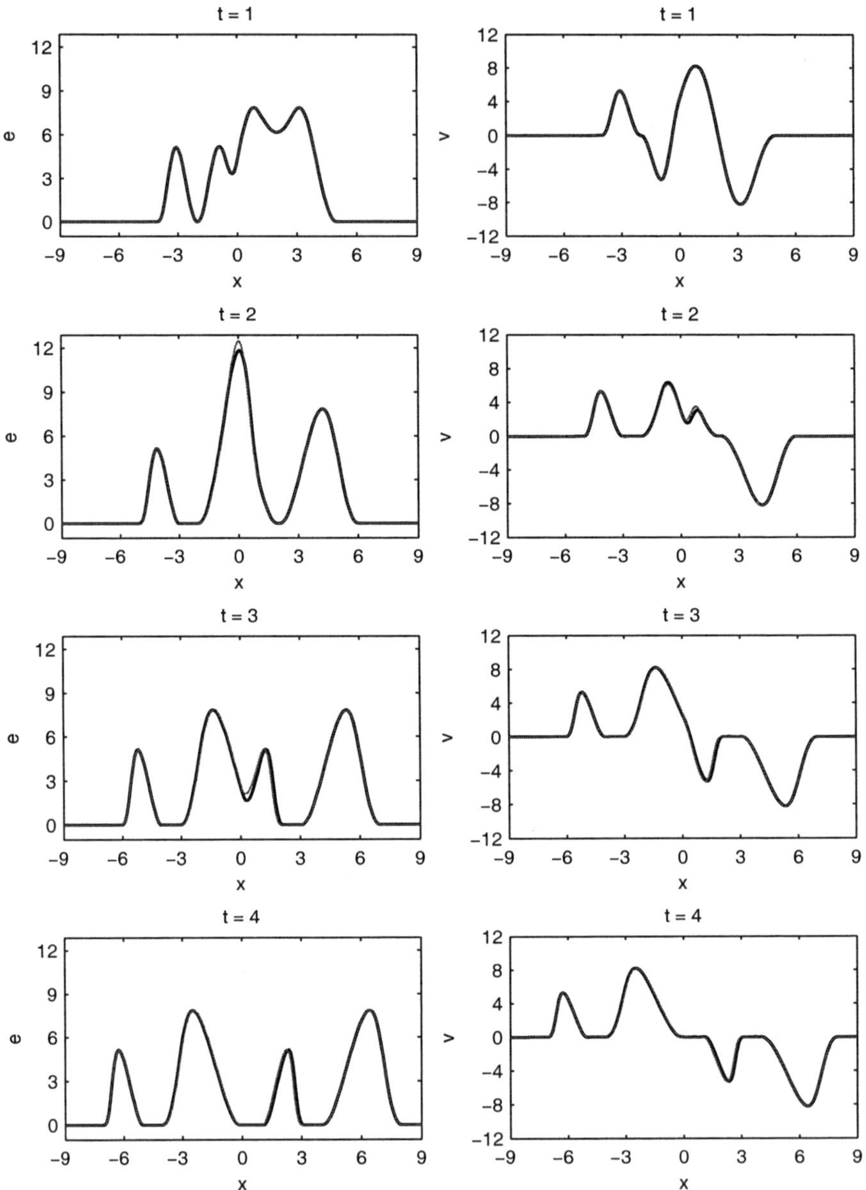

Fig. 3. Evolution of strain (left) and velocity (right) fields for initial data with two pulses (thick line), compared to the superposition of the evolutions for initial data with the two pulses taken separately (thin line), in the case $e_1 = 60$ under initial conditions shown in Figure 1(a).

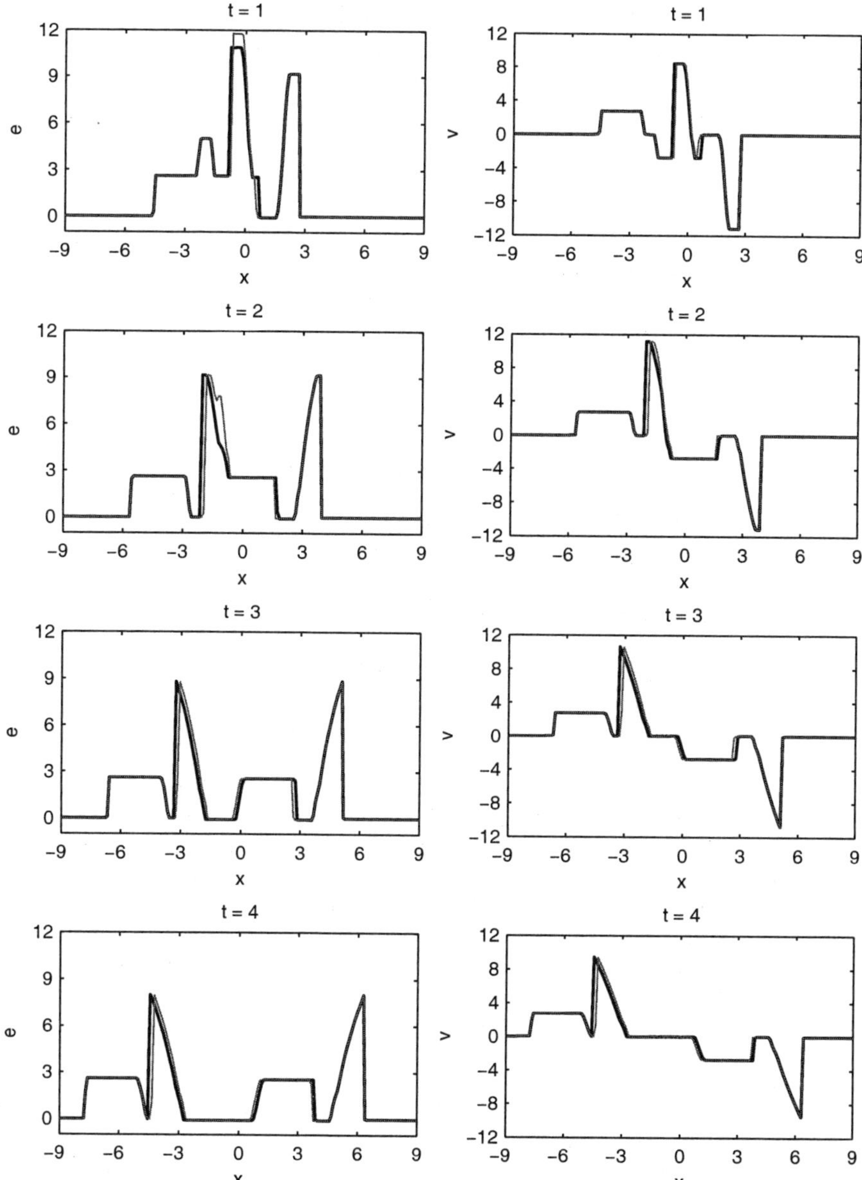

Fig. 4. Evolution of strain (left) and velocity (right) fields for initial data with two pulses (thick line), compared to the superposition of the evolutions for initial data with the two pulses taken separately (thin line), in the case $e_1 = 20$ under initial conditions shown in Figure 1(b).

compact support, such as those displayed in Figure 1, asymptotically converges to the solution of the corresponding zero Riemann problem. In the present situation, with and without interaction, the solution thus asymptotically converges to the null solution for the initial data considered here.

Acknowledgment. This paper was partially supported (A.M. and T.R.) by MIUR "Programma di Ricerca di Interesse Nazionale"(PRIN) *Non-linear Propagation and Stability in Thermodynamical Processes of Continuous Media* (Coordinator: T. Ruggeri) and by GNFM/INdAM " Progetto Giovani Ricercatori" (Coordinator: A. Mentrelli).

References

1. C. LOEWNER, A transformation theory of partial differential equations of gasdynamics, *Nat. Advis. Comm. Aeronaut. Tech. Notes* **2065**, 1-56 (1950).
2. G.A. DOMBROVSKII, *Approximation Methods in the Theory of Plane Adiabatic Gas Flow*, Moscow (1964).
3. D.L. CLEMENTS and C. ROGERS, On the theory of stress concentration for shear strained prismatical bodies with a nonlinear stress-strain law, *Mathematika* **22**, 34-42 (1975).
4. C. ROGERS, and W.F. SHADWICK, *Bäcklund Transformations and Their Applications*, Academic Press, New York (1982).
5. H.M. CEKIRGE and E. VARLEY, Large amplitude waves in bounded media: I. Reflexion and transmission of large amplitude shockless pulses at an interface, *Philos. Trans. Roy. Soc. London Ser A* **273**, 261-313 (1973).
6. J.Y. KAZAKIA and E. VARLEY, Large amplitude waves in bounded media: II. The deformation of an impulsively loaded slab: the first reflexion; III. The deformation of an impulsively loaded slab: the second reflexion, *Philos. Trans. Roy. Soc. London Ser A* **277**, 191-250 (1974).
7. B.R. SEYMOUR and E. VARLEY, Exact solutions describing soliton-like interactions in a non-dispensive medium, *SIAM J. Appl. Math.* **42**, 804-821 (1982).
8. C. ROGERS, W.K. SCHIEF and K.W. CHOW, On a novel class of model constitutive laws in nonlinear elasticity. Construction via Loewner theory, to appear in *Theor. Math. Phys.*.
9. T.W. DUERIG, A.R. PELTON and D. STÖCKEL, The use of superelasticity in medicine, *Metall* **50**, 569-574 (1996).
10. T.W. DUERIG, Present and future applictions of shape memory and superelastic materials, *Mat. Res. Soc. Symp.* **360**, 497-506 (1995).
11. C. ROGERS, W.K. SCHIEF and J. WYLIE, Wave propagation in ideally hard elastic inhomogeneous materials associated with pseudo-spherical surfaces, *Int. J. Eng. Sci.* **41**, 1965-1974 (2003).
12. W.K. SCHIEF and C. ROGERS, Loewner transformations. Adjoint and binary Darboux connections, *Stud. Appl. Math.* **100**, 391-422 (1998).

13. D. L. CLEMENTS and C. ROGERS, On the theory of shear concentration for shear-strained prismatic bodies with a nonlinear stress-strain law, *Mathematika* **22**, 34-42 (1975).

14. S. F. LIOTTA, V. ROMANO and G. RUSSO, Central scheme for balance laws of relaxation type, *SIAM J. Numer. Anal.* **38**, 1337-1356 (2000).

15. A. MENTRELLI and T. RUGGERI, Asymptotic behavior of Riemann and Riemann with structure problems for a 2×2 hyperbolic dissipative system, *Suppl. Rend. Circ. Mat. Palermo (Nonlinear hyperbolic fields and waves, A tribute to Guy Boillat)* **II/78**, 201-226 (2006).

16. T.-P. LIU, Linear and nonlinear large-time behavior of solutions of general systems of hyperbolic conservation laws, *Comm. Pure Appl. Math.* **30**, 767-796 (1977); Large-time behavior of solutions of initial and initial-boundary value problems of a general system of hyperbolic conservation laws, *Commun. Math. Phys.* **55**, 163-177 (1977).

FLAME STRUCTURE IN ORDINARY AND EXTENDED THERMODYNAMICS

INGO MÜLLER

Technical University Berlin

1. Introduction

A flame is similar to a shock wave. However, there are subtle diffenereces as follows:

- Flames propagate into the fuel, a gas in *metastable* chemical equilibrium so that an energetic barrier E must be overcome to *ignite* the fuel and make the flame possible.
- Flames are not adiabatic, because – once ignited – they set free the heat of reaction Q.

In a fairly simplistic manner the chemical reaction is modeled here by the stoichiometric formula $A \rightarrow B$, so that constituent A prevails before the flame and constituent B prevails behind the flame. Inside the flame both constituents are present. In this model there are only two constituents.

Fig. 1 represents a schematic picture of the fuel concentration and of the temperature. The *reaction zone* is the range where the heat of reaction is set free so that the temperature rises. The rise in temperature spreads out into the *preheating zone* – in the front – by heat conduction and thus makes the ignition possible. The inflow of mass – i.e. the speed of the flame – and the thermal conduction and the reaction rate must be delicately balanced, if the flame is to be stable: Too high an inflow will snuff out the flame, because the the fuel cannot all be burned so that the preheating will be insufficient. Too low an inflow will starve the flame, because the necessary preheating is prevented by a deficiency of chemical reaction.

Mathematically this delicate balance is reflected in the *flame eigenvalue μ* – a dimensionless combination of inflow, reaction rate and thermal conductivity.

Part of this paper is based on the theses [1] and [2] by M. Torrilhon and J. Au.

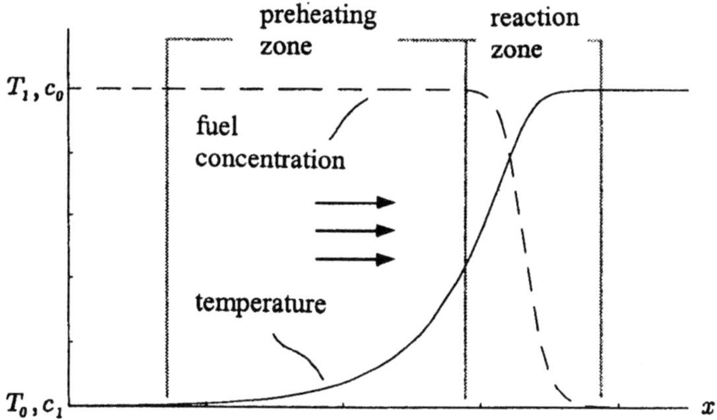

Fig. 1. Schematic view of a flame.

2. Equations of balance

The binary mixture considered here is governed by the equations of balance of the masses and momenta of the constituents and by the conservation law of energy of the mixture. Equivalently – using standard notation of continuum mechanics and thermodynamics – we may write these equations in the form.

$$\frac{\partial \rho}{\partial t} + \frac{\partial \rho v_i}{\partial x_i} = 0$$

$$\frac{\partial \rho v_j}{\partial t} + \frac{\partial (\rho v_j v_i - \sum_{\alpha=A}^{B}(t_{ij}{}^{\alpha} - \rho_{\alpha} u_j{}^{\alpha} u_i{}^{\alpha}))}{\partial x_i} = 0$$

$$\frac{\partial (\sum_{\alpha=A}^{B} \rho_{\alpha}(\varepsilon_{\alpha} + \frac{1}{2}u_{\alpha}{}^2) + \rho \frac{1}{2} v^2)}{\partial t} +$$
$$+ \frac{\partial (\sum_{\alpha=A}^{B} \rho_{\alpha}(\varepsilon_{\alpha} + \frac{1}{2}u_{\alpha}{}^2) + \rho \frac{1}{2} v^2)v_i - \sum_{\alpha=A}^{B}(t_{ij}{}^{\alpha} - \rho u_j{}^{\alpha} u_i{}^{\alpha})v_j + q_i{}^{\alpha} + \rho_{\alpha}(\varepsilon_{\alpha} + \frac{1}{2}u_{\alpha}{}^2)u_i{}^{\alpha} - t_{ij}{}^{\alpha} u_j^{\alpha})}{\partial x_i} = 0$$

$$\frac{\partial \rho c}{\partial t} + \frac{\partial (\rho c v_i + J_i)}{\partial x_i} = \tau$$

$$\frac{\partial (\rho c v_j + J_j)}{\partial t} + \frac{\partial (\rho c v_j v_i + v_j J_i + v_i J_j + \rho c u_i^1 u_j^1 - t_{ij}^1)}{\partial x_i} = m_i. \tag{2.1}$$

We neglect quadratic terms in the diffusion velocities u_i^{α} and ignore viscous stresses so that $t_{ij}{}^{\alpha} = -p^{\alpha}\delta_{ij}$ holds. c and J are concentration and diffusion flux of constituent A. Moreover we are interested in stationary and one-dimensional

146

conditions with $x_1 = x$ as the only independent variable. With all this the equations reduce to – with $h_\alpha = \varepsilon_\alpha + \dfrac{p_\alpha}{\rho_\alpha}$ as partial enthalpies

$$\frac{d\rho v}{dx} = 0$$

$$\frac{d(\rho v^2 + p)}{dx} = 0$$

$$\frac{d\left(\sum_{\alpha=A}^{B} h_\alpha(\rho_\alpha v + J_\alpha) + \rho \tfrac{1}{2} v^3 + \sum_{\alpha=A}^{B} q_\alpha\right)}{dx} = 0 \qquad \left(\sum_{\alpha=A}^{B} J_\alpha = 0\right) \tag{2.2}$$

$$\frac{d(\rho c v + J)}{dx} = \tau$$

$$\frac{d(\rho c v^2 + 2Jv + p_A)}{dx} = m_1.$$

Ordinary and extended thermodynamics differ in that the former theory ignores the last equation, i.e. the equation of balance of momentum of constituent A. This is tantamount to neglecting the inertia of the relative motion of the two constituents.

3.   Constitutive equations in ordinary thermodynamics

The system $(2.2)_{1\text{-}4}$ Of ordinary thermodynamics is not closed. It must be closed by constitutive equations for p, h_α, Σq_α, J, and τ.

We assume that both constituents are ideal gases with the same molecular mass m and the same constant specific heat c_p. In that case the thermal and caloric equations of state read

$$p = \rho \frac{k}{m} T \qquad \text{and} \qquad h_\alpha = h_\alpha^R + c_p(T - T_R), \tag{3.1}$$

where T_R is some reference temperature, usually chosen as 298K; $h_\alpha^{\ R}$ are the corresponding reference values of the enthalpies, they are constants.

The heat flux $\displaystyle\sum_{\alpha=A}^{B} q_\alpha$ and the diffusion flux J are given by the laws of Fourier and Fick respectively, viz.

$$\sum_{\alpha=A}^{B} q_\alpha = -\kappa \frac{dT}{dx} \qquad \text{and} \qquad J = -D \frac{dc}{dx}. \tag{3.2}$$

The mass production density τ of constituent A is equal to the reaction rate density λ to within a factor $-m$, and λ itself is proportional to the number density n_A so that we have

$$\tau = -m\lambda = -mKn_A = -\rho Kc.$$

The metastable character of the mixture before ignition is represented by the assumption

$$\rho K = a \exp(-\frac{E}{kT}),$$

where a is a constant and E is the activation energy. Thus in the simple theory the constitutive relation for the mass production density reads

$$\tau = -ac \exp[-\frac{E}{kT}]. \tag{3.3}$$

The value of E must be chosen such that τ is essentially zero far before the flame at the temperature T_o prevailing there.

4. Relevant equations of ordinary thermodynamics

We insert the constitutive equations into $(2.2)_{1\text{-}3}$ and integrate those equations from the state far ahead of the flame to some value x. Thus we obtain

$$\rho v = \rho_o v_o$$

$$\rho v^2 + \rho \frac{k}{m} T = \rho_o v_o^2 + \rho_o \frac{k}{m} T_o \tag{4.1}$$

$$-\kappa \frac{dT}{dx} + (h_A^R - h_B^R)(\rho c v + J) + \rho v c_p T + \rho \frac{1}{2} v^3 =$$
$$-\kappa \frac{dT}{dx}\Big|_o + (h_A^R - h_B^R)(\rho_o c_o v_o + J_o) + \rho_o v_o c_p T_o + \rho_o \frac{1}{2} v_o^{\,3}$$

$h_A^R - h_B^R$ is the reference value of the specific heat of reaction.

148

In addition we have $(2.2)_4$ and Fick's law, viz.

$$\frac{d(\rho c v + J)}{dx} = -ac \exp(-\frac{E}{kT})$$

$$J = -D\frac{dc}{dx}. \qquad (4.2)$$

[We do not insert Fick's law into $(4.1)_3$ and $(4.2)_1$ in order to avoid a second order differential equation.]

Given the values ρ_o, v_o, T_o, $c_o = 1$ and $\left.\frac{dT}{dx}\right|_o = 0$, $J_o = 0$ and given the constitutive parameters m, $h_A{}^R\text{-}h_B{}^R$, κ, c_p, a, E, and D we may solve this algebro-differential system so that in the state 1 far behind the flame $c_1 = 1$ holds. The solution is made more general by the introduction of dimensionless fields

$$\hat{\rho} = \frac{\rho}{\rho_o}, \quad \hat{v} = \frac{v}{v_o}, \quad \hat{T} = \frac{T}{T_o}, \quad \hat{j} = \frac{J}{\rho_o v_o}$$

and the dimensionless position variable

$$\hat{x} = \frac{c_p \rho_o v_o}{\kappa} x.$$

Also we introduce

- the dimensionless heat of reaction $Q = \dfrac{h_A^R - h_B^R}{c_p T_o}$,

- the dimensionless activation temperature $\hat{T}_{act} = \dfrac{E}{kT_o}$,

- the Mach number $M_o = \dfrac{v_o}{\sqrt{\gamma \frac{k}{m} T_o}}$, ($\gamma$ – ratio of specific heats)

- the dimensionless flame eigenvalue $\mu = \dfrac{a\kappa}{(\rho_o v_o)^2 c_p}$, and

- the dimensionless Lewis number defined as $Le = \dfrac{\kappa}{Dc_p}$.

The density $\hat{\rho}$ and the temperature $\hat{T}$ are both simply related to the velocity field $\hat{v}$ by $(4.1)_{1,2}$. We have

$$\hat{\rho} = \frac{1}{\hat{v}} \qquad \text{and} \qquad \hat{T} = (\gamma M_o^2 + 1)\hat{v} - \gamma M_o^2 \hat{v}^2 . \qquad (4.3)$$

Both fields may thus be eliminated from the system and we are left with three first order differential equations, viz.

$$[(\gamma M_o^2 + 1)\hat{v} - 2\gamma M_o^2 \hat{v}^2]\hat{v}'$$

$$= -\frac{\gamma+1}{2} M_o^2 + (\gamma M_o^2 + 1)\hat{v} + Q(c + J - 1) - (1 + \frac{\gamma-1}{2} M_0^2)$$

$$c' + J' = \mu c \exp(-\frac{\hat{T}_{act}}{(\gamma M_0^2 + 1)\hat{v} - \gamma M_o^2 \hat{v}^2})$$

$$c' = -cL . \qquad (4.4)$$

The prime denotes the derivative with respect to x.

5. Solution for ordinary thermodynamics. Chapman Jouguet relations

The most natural procedure for the solution of the equations (4.4) starts in state zero before the flame – with $c_o \approx 1$ – and proceeds by stepwise integration into and through the flame. The flame eigenvalue μ must be chosen as a shooting parameter so that behind the flame, in the state 1, the concentration assumes the value $c_1 = 0$.

This procedure, however, is quite impractical. Indeed, behind the flame our system of equations has a saddle point and it is never possible – by a numerical calculation – to reach a saddle point. The situation is quite similar mathematically to the calculation of a shock wave structure in a Navier-Stokes fluid with Fourier heat conduction, e.g. see [3]. And the resolution of the problem is also indicated by the procedure adopted in that case: *We need to shoot backwards from state 1 to state 0.*

A minor problem of the inversion of the integration is the *a priori* lack of knowledge of the velocity $\hat{v}_1$ behind the flame. However , that problem is easily solved. Indeed, both in front and behind the flame we expect c', J', and v' to be very close to zero and behind the flame we expect c and J to be essentially zero. Before the flame again J must be close to zero which is guaranteed by a small

150

value of the exponential expression in $(4.4)_2$. Thus (4.3) and $4.4)_1$ imply that the expressions

$$\hat{\rho}\hat{v} \qquad \text{and} \qquad \gamma M_o^2\hat{v} + \frac{\hat{T}}{\hat{v}} - (\gamma M_o^2 + 1)$$

$$\text{and} \qquad -\frac{\gamma+1}{2}M_0^2\hat{v}^2 + (\gamma M_0^2 + 1)\hat{v} + Q(c-1) - (1 + \frac{\gamma-1}{2}M_0^2)$$

must be equal far in front and far behind the flame. These are the Chapman Jouguet conditions[1] whose solutions read

$$\frac{\hat{v}_1}{\hat{v}_0} = \frac{\hat{\rho}_0}{\hat{\rho}_1} = 1 - \frac{M_o^2 - 1}{(\gamma+1)M_o^2}(1 + D(M_0,Q))$$

$$\frac{p_1}{p_o} = 1 + \frac{\gamma(M_o^2 - 1)}{\gamma+1}(1 + D(M_o^2,Q)), \qquad (5.1)$$

where

$$D = \pm\sqrt{1 - \frac{2M_o^2}{(M_o^2 - 1)^2}(\gamma+1)Q} \qquad \begin{array}{l} (+) - \text{det onation} \\ (-) - \text{flame} \end{array}$$

We choose the $-$ sign in (5.1), because we are interested in flames.

If we consider a problem with $\gamma = 4/3$, $M_o = 0.1$, $Q = 10$, the Chapman Jouguet solution (5.1) requires

$$\frac{\hat{v}_1}{\hat{v}_0} = 12.7196, \qquad \frac{\hat{p}_1}{\hat{p}_o} = 0.8437, \qquad \text{hence} \qquad \frac{\hat{T}_1}{\hat{T}_o} = 10.7320. \qquad (5.2)$$

Thus in particular, since $\hat{v}_o = 1$ holds, $\hat{v}_1$ is given by behind the flame and we may start the *backward integration* from

$$\bar{v}_1 = 12.7196, \qquad c_1 = 0.00001, \qquad \hat{J}_1 = 0.00001. \qquad (5.3)$$

For the actual calculation we choose parameters as follows:

$$\gamma = \frac{4}{3}, \qquad M_o = 0.1, \qquad Q = 10, \qquad L = 1, \qquad \hat{T}_{act} = 22 \qquad (5.4)$$

and an eigenvalue μ such that $c_o = 1$ holds.

---

[1]In ordinary shock structure analysis Q is zero and the Chapman Jouguet conditions are reduced to the Rankine Hugoniot conditions which are valid for shock waves.

In other words μ is used as a shooting parameter when we aim for $c_o = 1$. If μ is chosen as 31, c_o comes out too big; for $\mu = 25$, c_o comes out too small. Repeated shooting will provide the good value

$$\mu = 28.72$$

for which $c_o \approx 1$, see Fig. 2.

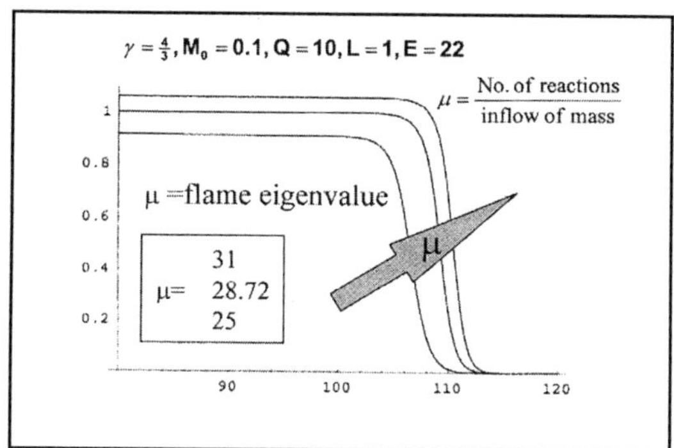

Fig. 2. On the choice of the flame eigenvalue μ.

All solutions $c(x)$, $\hat{v}(x)$, $\hat{J}(x)$, $\hat{T}(x,)$, and $\hat{p}(x)$ for the parameter values (5.4) and (5.5) are graphically shown in Fig. 3 by the solid lines.

Pressure and, of course, fuel concentration come down in a flame, while temperature and velocity go up. The diffusion flux is unequal to zero only inside the flame, because only there we have two constituents.

6. Relevant equations for extended thermodynamics

In the present case extended thermodynamics means no more than taking full account of the second momentum balance, viz. $(2.1)_5$ − or $(2.2)_5$ in the one-dimensional stationary case. This balance equation replaces Fick's law which does not occur in extended thermodynamics.

Everything that has been said and done before about the first four equations remains unchanged, but we need two more constitutive equations, namely for the partial pressure p_A and for the momentum production m_1 of constituent A.

Since we are dealing with two ideal gases with the same molecular mass we have

$$p_A = p\, c. \tag{6.1}$$

152

With this the momentum balance may fairly easily be cast into the form

$$v\frac{dJ}{dx} + 2J\frac{dv}{dx} + p\frac{dc}{dx} = m_1 - \tau v$$

of which the first two terms are due to the relative acceleration of the constituents A and B, terms which are neglected by ordinary thermodynamics. The rest of the equation must be compatible with Fick's law. From that requirement we conclude that the momentum production m_1 must have the form

$$m = \tau v - \frac{p}{D} J \tag{6.2}$$

Thus m_1 has two contributions, one due to the chemical reaction and one due to the relative motion of the constitutents. Both are eminently reasonable.

We use the same non-dimensional quantities as before and conclude that in extended thermodynamics Fick's equation (4.4) must be replaced by

$$\gamma M_o^2 \frac{\hat{v}}{\hat{T}}(\hat{v}J' + 2\hat{J}\hat{v}') + c' = -L\hat{J} \tag{6.3}$$

The other equations, namely (4.3), and $(4.4)_{1,2}$ remain unchanged.

So does the method of solution. For the final conditions (5.3) – behind the flame – and for the parameters (5.4) the differences between ordinary and extended thermodynamics are minimal but noticeable. The results of extended thermodynamics are graphically shown in Fig. 3 by the dashed curves. The flame eigenvalue for these curves is also minimally different from the value $\mu = 28.72$ of ordinary thermodynamics. Its new value is $\mu = 28.81$.

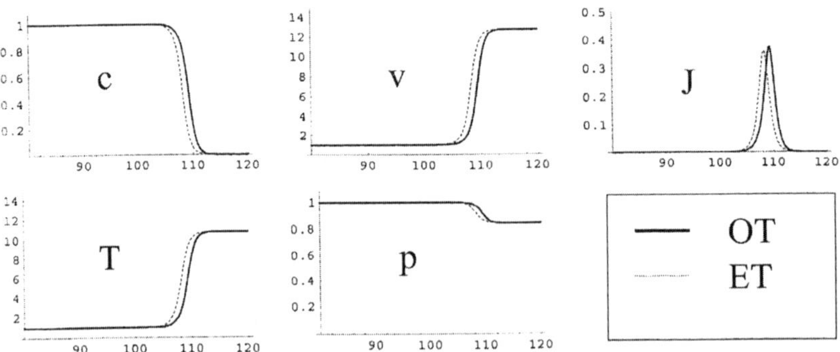

Solid: Ordinary thermodynamics. Dashed: Extended thermodynamics

Fig. 3. Fields in a flame.

7. References

[1] M. Torrilhon. Thermodynamische Berechnung der Ausbreitung von Flammen. Bachelor thesis TU Berlin (1999)

[2] J. D. Au. Lösung nichtlinearer Probleme in der Erweiterten Thermodynamik. Dissertation TU Berlin (2001)

[3] Gilbarg, D., Paolucci, D. The structure of shock waves in the continuum theory of fluids. J. Rat. Mecxh. Anal. 2, 1953

THE CHARACTERISTIC PROBLEM FOR THE EINSTEIN VACUUM EQUATIONS

Francesco Nicolò

Dipartimento di Matematica
Università degli Studi di Roma "Tor Vergata"
Via della Ricerca Scientifica, 00133-Roma, Italy

This is part of a talk given by the author at the conference held in Palermo (2006) in honor of Prof. Antonio Greco.

1. Introduction

The characteristic problem for the Einstein equations is the Cauchy problem with initial data given on a characteristic hypersurface. As in the Einstein equations the unknown function is the metric tensor, assigning the initial data for the characteristic problem means to specify a three dimensional manifold with a degenerate metric on it, together with some tensor fields which will acquire a specific geometrical meaning once the spacetime solution of the Einstein equations is obtained and the initial data hypersurface can be interpreted as an imbedded submanifold.

There are several reasons why the characteristic Cauchy problem for the Einstein equations deserves to be carefully investigated, some of them of "physical" nature, other of more "mathematical" flavour.

Between the first ones we observe that the characteristic hypersurfaces play a crucial role in General Relativity, describing the "null infinity"; moreover the horizons of the spacetimes with a black hole are also described by null hypersurfaces and, finally, data on a null cone describe the knowledge of the cosmos better than data given on a spacelike Cauchy hypersurface.

From the mathematical point of view we recall that, in general, the characteristic problem for the hyperbolic equations is somewhat more complicated that the non characteristic one due to the fact that the initial data cannot be assigned freely, but have to satisfy some constraints. Nevertheless in the case of the Einstein equations the situation is in a certain sense reversed. In this case, in fact, even the initial data of the non characteristic Cauchy problem cannot be assigned freely, but have to satisfy some constraints which are expressed through some elliptic equations involving the second fundamental form k of the initial spacelike hypersurface Σ_0 *.

There is a vast literature on the characteristic problem for linear and quasilinear hyperbolic equations and, more specifically, for the Einstein (vacuum) equations which, with due care, can be considered as belonging to the family of quasi linear hyperbolic equations[†]. The characteristic problem is naturally splitted in two separate problems: the problem of specifying initial data satisfying the constraint equations and the problem of the existence of the solution. This last problem is at its turn divide in two subproblems, namely the local existence and the global existence.

Let us concentrate from now on to the Einstein vacuum equations; in this case, the constraint problem has been discussed and solved by H.Muller Zum Hagen,,[Mu] H.Muller Zum Hagen and H.J.Seifert,[Mu-Se] and by A.Rendall,[Ren] when the hypersurface where the initial data are defined are the union of two transversely intersecting null hypersurfaces (H.Muller Zum Hagen and H.J.Seifert consider also the case where only one of the two hypersurfaces is null while the other one can be spacelike or timelike). The case, even more interesting, where the initial data satisfying the constraints are assigned on a "cone" is still an open problem at the level of the constraint problem (due to the fact that the constraints equations become singular at the vertex of the cone), see for instance the discussion on section 6 of,[Ren] while, assuming that the initial data constraints are satisfied, the existence has been proved by M.Dossa,[Do]

Concerning the existence problem, we note that in all these papers the local existence is proved[‡]. The strategy used in all of them is to consider the Einstein equations in the "harmonic" gauge. In[Mu-Se] H.Muller Zum Hagen and H.J.Seifert use more or less standard non linear hyperbolic technics, see for instance,[Br1] and their result is refined and completed in[Ch-Mu] and in.[Mu] The approach of A.Rendall[Ren] is somewhat different, he proves the existence and uniqueness of the (local) solution only for C^∞ initial data and does not discuss in detail the less regular case. Viceversa he focuses his attention on the explicit construction of initial data satisfying the harmonic condition and the constraint equations and produces a thorough examination of this problem.

Given this very schematic description of what is presently known about the characteristic problem for the Einstein equations, we can point two common aspects of all these approaches; the first one being that the problem is always considered in the harmonic gauge and the second that, with the exception of some result by M.Dossa,,[Do] these results are local existence results. Both aspects can be considered as drawbacks; the fact that the existence proof is local is an

---

*For a more exhaustive discussion on the relevance of the characteristic problem for the Einstein equations see,[Mu-Se Ren]

[†]The Einstein equations are in fact quasi linear hyperbolic equations if we make a definite choice of the "gauge", in particular the harmonic gauge. Nevertheless some general "hyperbolic" features are present even considering the Einstein equations in an arbitrary gauge.

[‡]The global existence is discussed by M.Dossa,[Do] but in a non linear hyperbolic case easier than the Einstein one.

obvious limitation, but we believe that to remain confined to the harmonic gauge is also a drawback as, in some sense, it is the origin of some difficulties. In fact the choice of a coordinate gauge does not allow to use, in all generality, the crucial geometric aspects of the Einstein equations[*].

After these considerations I want to discuss some results G.Caciotta and myself have recently obtained on this problem,[Ca-Ni1][Ca-Ni2] and some results still in progress.

More precisely we have studied the characteristic problem for the Einstein vacuum equations where the null initial data hypersurface is the union of two portions of (truncated) null cones one outgoing, the second one incoming, having their vertices on the same vertical axis, the vertex of the incoming cone being above the one of the outgoing one. The two cones, we call hereafter C_0 and $\underline{C}_0$, intersect along a two dimensional S^2 surface hereafter denoted S_0. Looking at these cones as null cones of the Minkowski spacetime we can imagine S_0 as lying on the spacelike hyperplane $t = 0$. It is important to remark that this is an intuitive and non rigorous description of the null hypersurface $\mathcal{C} = C_0 \cup \underline{C}_0$ of the characteristic problem, as, in fact, C_0 and $\underline{C}_0$ are not exactly cones, but simply null hypersurfaces as no spherical symmetry is imposed, their "vertices" lie only approximately on the same vertical axis and $S_0 = C_0 \cap \underline{C}_0$ is simply S^2 diffeomorphic. On the other side, as we want to solve an existence problem with small initial data and as the data given on $\mathcal{C}$ describe its geometric properties this allows us to conclude that this picture is, in a sense which can be made precise, very near to the real one. These null hypersurfaces C_0 and $\underline{C}_0$ are in fact small deformations of exact Minkowski cones and, analogously S_0 is very near to an S^2 surface. It is also appropriate to observe that, nevertheless, we can, proceeding as in the non characteristic case, define the initial data of our problem, namely the null hypersurface $\mathcal{C}$ and some tensor fields defined on it, in a completely intrinsic way, see for this aspect Section 1 of.[Ca-Ni1]

The problem we have considered belongs, therefore, to the class of problems considered in,[Mu-Se][Ren] In fact in those papers the characteristic hypersurface is the union of two general null hypersurfaces and, although the authors seem to consider as hypersurfaces approximate null hyperplanes, nevertheless the case where they are (approximate) portion of cones seems suitable to be treated in the same way.

Once defined the problem we want to solve, let me describe the more significant differences with the previous results. They are essentially of two types, associated to the two different problems in which a characteristic non linear problem naturally splits: the constraints problem and the existence problem.

a) Concerning the constraints problem, we do not choose the "harmonic gauge" to define the initial data. This allows us to use an approach basically coordinate independent, the gauge we use being simply defined by the choice of a specific class of foliations of $C_0 \cup \underline{C}_0$, a procedure which allows us to use the full

---

[*]Nevertheless a more refined use of the harmonic gauge made by H.Lindblad and I.Rodnianski, see,[Ro-Li] has allowed to obtain also in this gauge a global result for the non characteristic problem.

richness of the Einstein equations geometric structure. Moreover, proceeding in this way, the constraint problem is disentangled in a clear fashion from the gauge choice, which was not true in the previous approaches as can be seen in particular in the A.Rendall paper,[Ren] where the harmonic conditions and the constraints conditions have to be satisfied together.

b) Concerning the existence problem as we already said we have to prove first a local existence result and later on the global one. We have, nevertheless, inverted the order using the fact that, in principle, for the local result we could use, paying some attention, the existing local existence proofs*. Therefore, assuming a local existence result given, we have proved first a global result. This is significantly based on a result that S.Klainerman and myself,,[Kl-Ni2] have obtained some years ago about the global existence (for an "external" region) of a non characteristic Cauchy problem with initial data asymptotically flat and small. Using analogous technics, suitably adapted to the characteristic problem, G.Caciotta and myself have proved, also in this case, a global existence result for a spacetime with complete null outgoing geodesics.

c) The third part of our result is a new proof of the local existence. This work is still in progress and some technical details have to be completed. Again the main idea is to prove the local existence without using the harmonic gauge which means exploiting the intrinsic hyperbolicity of the Einstein equations. The solution is obtained as limit of a sequence of real analytic solutions obtained via Cauchy-Kowalevski theorem and the crucial step is the use of the Bel-Robinson norms as energy-type norms respect to which the sequence of real analytic solutions tend to a solution in a appropriate Sobolev space.

Once described the main steps of this result, let me quote also its limits which shed also some light on the remaining open problems.

The first thing to remark is that, as $\mathcal{C}$ is the union of an outgoing cone and a portion of an incoming cone, its development, whose we prove the existence, is an "external" region in the sense that, thinking to these cones as immersed in a R^4 manifold, the maximal development does not contain the origin nor any point belonging to the origin influence region. This is due to the fact that the strategy for the proof is borrowed from the one used in[Kl-Ni2] where, again, an external problem was studied. The difficulty in matching the external problem solution with an internal one is connected to the difficulty of solving the characteristic problem with initial data on a null "cone", a problem still open and whose difficulty resides mainly in the way of treating the constraint equations near the tip of the cone.

The second limit of our result is that it is a small data result. Of course if the data were not small we would not expect that, in general, the outgoing null geodesics be complete, as a consequence of the singularity theorems of R.Penrose and S.W.Hawking, see[Haw-El] for a review. Nevertheless being able to treat large initial data would allow to face the important problem of the dynamical origin of a singularity, a task extremely difficult, but of great interest*.

Concerning the first limit, we expect that the "internal" characteristic result

---

*The attention is needed to adapt our initial data to the harmonic gauge.
*see also on this subject.[Ch2]

can be treated in the same way we solve the present problem when, instead of considering $\mathcal{C}$ the union of an outgoing cone and an incoming one, we consider $\mathcal{C}$ as a single cone with vertex on the timeline of the origin for a negative time, truncated by the portion of the hyperplane $t = 0$ intersecting it. We believe this result can be obtained basically within this approach, but it will present some extra difficulties.

2. The main technical aspects

In this section, without giving too many details, I point out the more relevant aspects of the proof. Again it is natural to treat separately the costraint problem, the local existence problem and the global existence problem.

2.1. *The constraint problem*

2.1.1. *The definition of the initial data*

In[Kl-Ni2] one of the main tool to obtain the global existence result is the foliation of the spacetime with two families of null hypersurfaces (the equivalent of cones in the Minkowski spacetime) outgoing and incoming respectively.

This double foliation made by two families of null hypersurfaces, we denote $\{C(\lambda)\}$ and $\{\underline{C}(\nu)\}$, is such that each intersection $S(\lambda, \nu) = C(\lambda) \cap \underline{C}(\nu)$ is a two dimensional surface diffeomorphic to S^2. Associated to this foliation we can define a null moving frame "adapted" to it, $\{e_4, e_3 . e_A\}$, where e_4 and e_3 are two vector fields proportional to the tangent vector fields L and $\underline{L}$ of the null geodesics generating the null cones C_0 and $\underline{C}_0$, $e_4 = 2\Omega L, e_3 = 2\Omega \underline{L}$ and $\{e_A\}$, $A \in \{1, 2\}$ is an orthonormal frame adapted to the $S(\lambda, \nu)$ surfaces.

Once the moving frame is assigned in $\mathcal{M}$ it is possible to define the Ricci coefficients (connection coefficients), and look at the structure equations they satisfy. The structure equations describe the way the null frame changes from place to place in $\mathcal{M}^*$. More explicitly the connection coefficients associated to a "specific"[†] null moving frame are:

$$\chi_{ab} = \mathbf{g}(\mathbf{D}_{e_a} e_4, e_b) \quad , \quad \underline{\chi}_{ab} = \mathbf{g}(\mathbf{D}_{e_a} e_3, e_b)$$

$$\underline{\xi}_a = \frac{1}{2}\mathbf{g}(\mathbf{D}_{e_3} e_3, e_a) = \frac{1}{2}e_3(\log \Omega)\mathbf{g}(e_3, e_a) = 0$$

$$\xi_a = \frac{1}{2}\mathbf{g}(\mathbf{D}_{e_4} e_4, e_a) = \frac{1}{2}e_4(\log \Omega)\mathbf{g}(e_4, e_a) = 0$$

$$\underline{\omega} = \frac{1}{4}\mathbf{g}(\mathbf{D}_{e_3} e_3, e_4) = -\frac{1}{2}\mathbf{D}_3(\log \Omega)$$

$$\omega = \frac{1}{4}\mathbf{g}(\mathbf{D}_{e_4} e_4, e_3) = -\frac{1}{2}\mathbf{D}_4(\log \Omega) \tag{1}$$

$$\underline{\eta}_a = -\zeta_a + \nabla\!\!\!/_a \log \Omega \ , \ \eta_a = \zeta_a + \nabla\!\!\!/_a \log \Omega$$

$$\zeta_a = \frac{1}{2}\mathbf{g}(\mathbf{D}_{e_a} e_4, e_3) \ .$$

---

[*]see for instance,[Sp] Vol 2.
[†]See for details,[Kl-Ni2] Chapter 3.

The first set of the structure equations is *

$$\mathbf{D}_a e_b = \mathbf{\nabla}_a e_b + \frac{1}{2}\chi_{ab}e_3 + \frac{1}{2}\underline{\chi}_{ab}e_4$$

$$\mathbf{D}_a e_3 = \underline{\chi}_{ab}e_b + \zeta_a e_3 \quad , \quad \mathbf{D}_a e_4 = \chi_{ab}e_b - \zeta_a e_4$$

$$\mathbf{D}_3 e_a = \mathbf{\not{D}}_3 e_a + \eta_a e_3 \quad , \quad \mathbf{D}_4 e_a = \mathbf{\not{D}}_4 e_a + \underline{\eta}_a e_4 \tag{2}$$

$$\mathbf{D}_3 e_3 = (\mathbf{D}_3 \log\Omega)e_3 \quad , \quad \mathbf{D}_3 e_4 = -(\mathbf{D}_3 \log\Omega)e_4 + 2\eta_b e_b$$

$$\mathbf{D}_4 e_4 = (\mathbf{D}_4 \log\Omega)e_4 \quad , \quad \mathbf{D}_4 e_3 = -(\mathbf{D}_4 \log\Omega)e_3 + 2\underline{\eta}_b e_b \;.$$

The second set of the structure equations connect the first derivatives of the connection coefficients to the conformal part of the curvature tensor and to the Ricci tensor of the spacetime. Those which depend on the Ricci tensor, denoting $\mathbf{R}_{\alpha,\beta} = \mathbf{Ricci}(e_\alpha, e_\beta)$, are †

$$\mathbf{D}_4 \mathrm{tr}\chi + \frac{1}{2}(\mathrm{tr}\chi)^2 + 2\omega\mathrm{tr}\chi + |\hat\chi|^2 = \mathbf{R}_{44}$$

$$\mathbf{D}_3 \mathrm{tr}\underline{\chi} + \frac{1}{2}(\mathrm{tr}\underline{\chi})^2 + 2\underline{\omega}\mathrm{tr}\underline{\chi} + |\underline{\hat\chi}|^2 = \mathbf{R}_{33}$$

$$\mathbf{\not{D}}_4 \zeta + \frac{3}{2}\mathrm{tr}\chi\zeta + \zeta\hat\chi - \mathrm{d\!\!/v}\,\hat\chi + \frac{1}{2}\mathbf{\not\nabla}\mathrm{tr}\chi + \mathbf{\not{D}}_4\mathbf{\not\nabla}\log\Omega = \mathbf{R}_{4a} \tag{3}$$

$$\mathbf{\not{D}}_3 \zeta + \frac{3}{2}\mathrm{tr}\underline{\chi}\zeta + \zeta\underline{\hat\chi} + \mathrm{d\!\!/v}\,\underline{\hat\chi} - \frac{1}{2}\mathbf{\not\nabla}\mathrm{tr}\underline{\chi} - \mathbf{\not{D}}_3\mathbf{\not\nabla}\log\Omega = \mathbf{R}_{3a}$$

$$\mathbf{\not{D}}_4 \mathrm{tr}\underline{\chi} + \mathrm{tr}\chi\mathrm{tr}\underline{\chi} - 2\omega\mathrm{tr}\underline{\chi} - 2(\mathrm{d\!\!/v}\,\underline{\eta}) - 2|\underline{\eta}|^2 + 2\mathbf{K} = \mathrm{tr}\mathbf{R}$$

$$\mathbf{\not{D}}_4 \underline{\hat\chi}_{ab} + \frac{1}{2}\mathrm{tr}\chi\underline{\hat\chi}_{ab} + \frac{1}{2}\mathrm{tr}\underline{\chi}\hat\chi_{ab} - 2\omega\underline{\hat\chi}_{ab} - 2(\mathbf{\not\nabla}\widehat\otimes\underline{\eta})_{ab} - 2(\underline{\eta}\widehat\otimes\underline{\eta})_{ab} = \hat{\mathbf{R}}_{ab}$$

$$\mathbf{\not{D}}_3 \mathrm{tr}\chi + \mathrm{tr}\underline{\chi}\mathrm{tr}\chi - 2\underline{\omega}\mathrm{tr}\chi - 2(\mathrm{d\!\!/v}\,\eta) - 2|\eta|^2 + 2\mathbf{K} = \mathrm{tr}\mathbf{R}$$

$$\mathbf{\not{D}}_3 \hat\chi_{ab} + \frac{1}{2}\mathrm{tr}\underline{\chi}\hat\chi_{ab} + \frac{1}{2}\mathrm{tr}\chi\underline{\hat\chi}_{ab} - 2\underline{\omega}\hat\chi_{ab} - 2(\mathbf{\not\nabla}\widehat\otimes\eta)_{ab} - 2(\eta\widehat\otimes\eta)_{ab} = \hat{\mathbf{R}}_{ab}$$

$$\mathbf{D}_3 \omega + \mathbf{D}_4 \underline{\omega} - 4\omega\underline{\omega} - 3|\zeta|^2 + |\mathbf{\not\nabla}\log\Omega|^2 - \frac{1}{2}\bigg[\mathbf{D}_4\mathrm{tr}\underline{\chi} + \frac{1}{2}\mathrm{tr}\chi\mathrm{tr}\underline{\chi} - 2\omega\mathrm{tr}\underline{\chi}$$

$$+\hat\chi\cdot\underline{\hat\chi} - 2\mathrm{d\!\!/v}\,\underline{\eta} - 2|\underline{\eta}|^2\bigg] = \mathbf{R}_{34} \;.$$

The remaining structure equations of the second group depend on the conformal part of the Riemann tensor, $\mathbf{C}$, decomposed with respect to the adapted null

---

* Hereafter $\mathbf{D}_{e_a} = \mathbf{D}_a$ and $\mathbf{D}_{e_{(3,4)}} = \mathbf{D}_{(3,4)}$, $\mathbf{\not\nabla}$ is the connection associated to the metric restriction on $S(\lambda,\nu)$, γ_{ab}, and $\mathbf{\not{D}}_{3,4}F$ is the projection of the covariant derivative of F along the null direction $e_{3,4}$ on the tangent plane of $S(\lambda,\nu)$.

† $2^{-1}{}^{(2)}\mathbf{R} = \mathbf{K}$, the scalar curvature of the leaves $S(\lambda,\nu)$, and $\hat{\mathbf{R}}_{ab}$ is the traceless part of $\mathbf{R}_{ab}$, with respect to the a, b indices.

frame, they are: * †

$$\not{D}_4\hat{\chi} + \mathrm{tr}\chi\hat{\chi} - (\mathbf{D}_4\log\Omega)\hat{\chi} = -\alpha$$

$$\not{D}_4\zeta + 2\chi\cdot\zeta + \not{D}_4\not{\nabla}\log\Omega = -\beta$$

$$\not{\nabla}\mathrm{tr}\chi - \not{\mathrm{div}}\chi - \zeta\cdot\chi + \zeta\mathrm{tr}\chi = \beta$$

$$\mathbf{D}_4\mathrm{tr}\underline{\chi} + \frac{1}{2}\mathrm{tr}\chi\mathrm{tr}\underline{\chi} + (\mathbf{D}_4\log\Omega)\mathrm{tr}\underline{\chi} + \hat{\chi}\cdot\hat{\underline{\chi}} + 2\not{\mathrm{div}}\zeta - 2\not{\triangle}\log\Omega$$

$$-2|\zeta|^2 - 4\zeta\cdot\not{\nabla}\log\Omega - 2|\not{\nabla}\log\Omega|^2 = 2\rho \tag{5}$$

$$\mathbf{D}_3\mathrm{tr}\chi + \frac{1}{2}\mathrm{tr}\underline{\chi}\mathrm{tr}\chi + (\mathbf{D}_3\log\Omega)\mathrm{tr}\chi + \hat{\underline{\chi}}\cdot\hat{\chi} - 2\not{\mathrm{div}}\zeta - 2|\zeta|^2$$

$$-2\not{\triangle}\log\Omega - 4\zeta\cdot\not{\nabla}\log\Omega - 2|\not{\nabla}\log\Omega|^2 = 2\rho$$

$$\mathbf{K} + \frac{1}{4}\mathrm{tr}\chi\mathrm{tr}\underline{\chi} - \frac{1}{2}\hat{\underline{\chi}}\cdot\hat{\chi} = -\rho$$

$$\not{\mathrm{curl}}\zeta - \frac{1}{2}\hat{\chi}\wedge\hat{\underline{\chi}} = \sigma$$

$$\not{D}_3\zeta + 2\underline{\chi}\cdot\zeta - \not{D}_3\not{\nabla}\log\Omega = -\underline{\beta}$$

$$\not{\nabla}\mathrm{tr}\underline{\chi} - \not{\mathrm{div}}\underline{\chi} + \zeta\cdot\underline{\chi} - \zeta\mathrm{tr}\underline{\chi} = -\underline{\beta}$$

$$\not{D}_3\hat{\underline{\chi}} + \mathrm{tr}\underline{\chi}\,\hat{\underline{\chi}} - (\mathbf{D}_3\log\Omega)\hat{\underline{\chi}} = -\underline{\alpha}$$

If we consider these equations in a vacuum Einstein spacetime, then equations 3 can be seen as a way of writing the vacuum Einstein equations in terms of the connection coefficients and their first derivatives.‡

To pose the characteristic problem for the vacuum Einstein equations, we write the initial data given on C in terms of the metric tensor components and the connection coefficients restricted to the initial hypersurface. Therefore the subset 3 of the structure equations will play the role of constraint equations.

Definition 2.1. The initial data set of the characteristic problem consists in an initial data set relative to the "null outgoing cone" and an initial data set relative

---

*

$$\alpha(X,Y) = \mathbf{C}(X,e_4,Y,e_4) \ , \quad \underline{\alpha}(X,Y) = \mathbf{C}(X,e_3,Y,e_3)$$

$$\beta(X) = \frac{1}{2}\mathbf{C}(X,e_4,e_3,e_4) \ , \quad \underline{\beta}(X) = \frac{1}{2}\mathbf{C}(X,e_3,e_3,e_4) \tag{4}$$

$$\rho = \frac{1}{4}\mathbf{C}(e_3,e_4,e_3,e_4) \ , \quad \sigma = \frac{1}{4}{}^{\star}\mathbf{C}(e_3,e_4,e_3,e_4)$$

with X,Y vector fields tangent to $S(\lambda,\nu)$ and ${}^{\star}\mathbf{C}$ the left Hodge dual of $\mathbf{C}$.

†The structure equations are automatically satisfied in any Lorentzian manifold, $(\mathcal{M},\mathbf{g})$. They can be interpreted as a way of rewriting the first order covariant derivatives of the moving frame in terms of the connection coefficients, see equations 2. In the same way equations, 3 and 5, express the Ricci part and the conformal part of the Riemann tensor in terms of the connection coefficients and their first derivatives.

‡This set of equations is larger than what is needed. In fact looking at 3 we see that the equations associated to $\hat{\mathbf{R}}_{ab} = 0$ and to $\mathrm{tr}\mathbf{R} = 0$ appear repeated twice.

to the "null incoming cone" *,

$$\{C_0 \; ; \gamma_{ab}, \Omega, \zeta, \underline{\chi}, \underline{\omega}\} \cup \{\underline{C}_0 \; ; \gamma_{ab}, \underline{\Omega}, \zeta, \chi, \omega\} \; . \tag{6}$$

On C_0, $\{\underline{\chi}, \underline{\omega} \, , \zeta, \chi_{ab} = \frac{1}{2\Omega} \frac{\partial \gamma_{ab}}{\partial \nu}\}$ satisfy the "constraint equations"[†]:

$$\frac{\partial tr\chi}{\partial \nu} + \frac{\Omega tr\chi}{2} tr\chi + 2\Omega\omega tr\chi + \Omega|\hat{\chi}|^2 = 0$$

$$\frac{\partial \zeta_A}{\partial \nu} + \Omega\chi_{AB}\zeta_B + \Omega tr\chi\zeta_A - \Omega(\text{div}\chi)_A + \Omega\nabla\!\!\!\!/_A tr\chi + \frac{\partial \nabla\!\!\!\!/_A \log \Omega}{\partial \nu} = 0$$

$$\frac{\partial tr\underline{\chi}}{\partial \nu} + \Omega tr\chi tr\underline{\chi} - 2\Omega\omega tr\underline{\chi} - 2\Omega(\text{div}\underline{\eta}) - 2\Omega|\underline{\eta}|^2 + 2\Omega\mathbf{K} = 0$$

$$\frac{\partial \hat{\underline{\chi}}_{AB}}{\partial \nu} + \frac{\Omega tr\chi}{2}\hat{\underline{\chi}}_{AB} + \frac{\Omega tr\underline{\chi}}{2}\hat{\chi}_{AB} + (\frac{\partial \log \Omega}{\partial \nu})\hat{\underline{\chi}}_{AB} + \Omega\nabla\!\!\!\!/_A \widehat{\otimes}(\zeta - \nabla\!\!\!\!/ \log \Omega)_B$$

$$-\Omega(\zeta_A - \nabla\!\!\!\!/_A \log \Omega)\widehat{\otimes}(\zeta - \nabla\!\!\!\!/ \log \Omega)_B = 0 \tag{7}$$

$$\frac{\partial \underline{\omega}}{\partial \nu} - 2\Omega\omega\underline{\omega} - \Omega\zeta_C\nabla\!\!\!\!/_C \log \Omega + \frac{1}{2}\Big[-3\Omega|\zeta|^2 + \Omega|\nabla\!\!\!\!/ \log \Omega|^2$$

$$+\Big(\mathbf{K} + \frac{1}{4}tr\chi tr\underline{\chi} - \frac{1}{2}\hat{\chi} \cdot \hat{\underline{\chi}}\Big)\Big] = 0$$

---

*We denote $\{S_0(\nu)\}$ and $\{\underline{S}_0(\lambda)\}$ the leaves of the foliations of C_0 and $\underline{C}_0$ such that $\underline{S}(\lambda_1) = S(\nu_0) = S_0 = C_0 \cap \underline{C}_0.$, $S_0(\nu) = \{p \in C_0 | \underline{u}(p) = \nu\}$, $\underline{S}_0(\lambda) = \{p \in \underline{C}_0 | u(p) = \lambda\}$, $\nu \in [\nu_0, \infty), \lambda \in [\lambda_1, \lambda_0 < 0]$. The functions $\underline{u}(p)$ and $u(p)$ are defined in the following way: let $\gamma(\underline{v})$ be the null geodesic on C_0 starting on S_0 with affine parameter $\underline{v} = 0$ and passing through p when $\underline{v} = \underline{v}(p)$, then $\underline{u}(p) = \nu_0 + \int_0^{\underline{v}(p)} (4\Omega)^{-2}(\gamma(\underline{v}'))d\underline{v}'$, in the same way $u(p) = \lambda_1 + \int_0^{v(p)} (4\underline{\Omega})^{-2}(\gamma(v'))dv'$, the scalar function Ω being the null lapse function for the metric on $\mathcal{C}$.

[†] γ_{ab} is the restriction of the initial data metric on the leaves $\{S_0(\nu)\}$ and $\{\underline{S}_0(\lambda)\}$ of $\mathcal{C}$. Observe also that on each cone there are seven constraint equations while we expect the correct number be six. In fact, for C_0, the equation associated to $\mathbf{R}_{44}$ (respectively $\mathbf{R}_{33}$ for $\underline{C}_0$) has to be considered connected to the gauge choice.

162

and on $\underline{C}_0$, $\{\chi, \omega, \zeta,\ \underline{\chi}_{ab} = \frac{1}{2\Omega}\frac{\partial \gamma_{ab}}{\partial \lambda}\}$ satisfy the "constraint equations"[*]:

$$\frac{\partial tr\underline{\chi}}{\partial \lambda} + \frac{\Omega tr\underline{\chi}}{2} tr\underline{\chi} + 2\Omega\omega\, tr\underline{\chi} + \Omega|\hat{\underline{\chi}}|^2 = 0$$

$$\frac{\partial \zeta_A}{\partial \lambda} + \Omega\underline{\chi}_{AB}\zeta_B + \Omega tr\underline{\chi}\zeta_A + \Omega(\not\!div\underline{\chi})_A - \Omega\not\!\nabla_A tr\underline{\chi} - \frac{\partial \not\!\nabla_A \log\Omega}{\partial \lambda} = 0$$

$$\frac{\partial tr\chi}{\partial \lambda} + \Omega tr\underline{\chi}\, tr\chi - 2\Omega\underline{\omega}\, tr\chi - 2\Omega(\not\!div\eta) - 2\Omega|\eta|^2 + 2\Omega\mathbf{K} = 0$$

$$\frac{\partial \hat{\chi}_{AB}}{\partial \lambda} + \frac{\Omega tr\underline{\chi}}{2}\hat{\chi}_{AB} + \frac{\Omega tr\chi}{2}\hat{\underline{\chi}}_{AB} + \frac{\partial \log\Omega}{\partial \lambda}\hat{\chi}_{AB} - \Omega\not\!\nabla_A\widehat{\otimes}(\zeta + \not\!\nabla\log\Omega)_B$$

$$-\Omega(\zeta_A + \not\!\nabla_A\log\Omega)\widehat{\otimes}(\zeta + \not\!\nabla\log\Omega)_B = 0 \tag{8}$$

$$\frac{\partial \omega}{\partial \lambda} - 2\Omega\omega\underline{\omega} + \Omega\zeta_C\not\!\nabla_C\log\Omega + \frac{1}{2}\left[-3\Omega|\zeta|^2 + \Omega|\not\!\nabla\log\Omega|^2\right.$$

$$\left. +\Omega\left(\mathbf{K} + \frac{1}{4}tr\chi\, tr\underline{\chi} - \frac{1}{2}\hat{\chi}\cdot\hat{\underline{\chi}}\right)\right] = 0\ .$$

Given Definition 2.1 we can state precisely the characteristic initial value problem as the construction of an Einstein spacetime $\{\mathcal{M}, \mathbf{g}\}$ and of an imbedding i such that the the null hypersurface $\mathcal{C}$ of the initial data becomes the union of two null hypersurfaces of $\mathcal{M}$ and the tensor quantities defined as initial data acquire a geometrical meaning as the pull-back of the connection coefficients of $\mathcal{M}$ associated to an adapted null frame. A more detailed statement of the characteristic initial value problem is given in.[Ca-Ni1]

2.1.2. *The construction of the initial data*

Once we have defined the constraint equations that the initial data have to satisfy the next goal is to prove that initial data satisfying them can be obtained. This problem is somewhat analogous to the problem of solving the constraint equations

$$\nabla^j k_{ij} - \nabla_i \mathrm{tr} k = 0\ ,\quad R - |k|^2 + (\mathrm{tr}k)^2 = 0\ . \tag{9}$$

of the non characteristic problem. Moreover, as done for the problem 9, see,[Ch-Kl],[Kl-Ni2] while trying to solve the characteristic constraint problem we will also prescribe the initial data decay along the null hypersurfaces C_0 and $\underline{C}_0$ when their coordinate ν or $|\lambda|$ tend to ∞. The decay rate we require is such that

---

[*]a) Although the definitions of the initial data and the constraint equations they have to satisfy are borrowed by the structure equations for the connection coefficients in a generic Lorentzian manifold it is worth to emphasize that, formally, the tensors $\chi, \zeta, \underline{\chi}\cdots$ will acquire their geometrical meaning while proving the existence result and the initial data hypersurface $\mathcal{C}$ is interpreted as an imbedded hypersurface of the vacuum spacetime $\mathcal{M}$.

b) Obviously, to prove the existence result, the initial data have to satisfy, together with the previous constraint equations, some regularity conditions. We will discuss it in more detail when we consider the existence problem.

some "energy type" integral norms defined on C_0 and $\underline{C}_0$ are bounded. In fact to apply the techniques used in[Kl-Ni2] to prove our global existence result, we need that some "flux integrals" are bounded. In the non characteristic case this implies that these integral norms, $\mathcal{Q}$, see their precise definition in,[Kl-Ni2] Chapter 3, are bounded in terms of the corresponding (finite) norms defined on the initial hypersurface Σ_0. In the characteristic case the $\mathcal{Q}$ flux norms have to be bounded in terms of the same flux norms defined on the initial hypersurface C. This requires that the various null (conformal) Riemann components decay sufficiently fast along the initial cones. In particular, if the $\mathcal{Q}$ norms we use are those defined in,[Kl-Ni2] it follows that on C_0 and $\underline{C}_0$ we need at least the following decay, with $\delta > 0$:

$$\alpha = O(r^{-(\frac{7}{2}+\delta)}) \, , \quad \beta = O(r^{-(\frac{7}{2}+\delta)}) \, , \quad (\rho - \overline{\rho}, \sigma) = O(r^{-3}|\lambda|^{-(\frac{1}{2}+\delta)})$$

$$\underline{\beta} = O(r^{-2}|\lambda|^{-(\frac{3}{2}+\delta)}) \, , \quad \underline{\alpha} = O(r^{-1}|\lambda|^{-(\frac{5}{2}+\delta)}) \, . \tag{10}$$

We do not give here the long and detailed proof of this result and we refer to[Ca-Ni1] for all the technical aspects.

Once the constraint equations for the characteristic problem has been precisely defined and solved we can face the existence problem. As said before we assume that a local solution is at our disposal and prove first a global existence result.

2.2. *The global existence problem*

Once the initial data satisfying the constraints are assigned with the appropriate decay and regularity the global result can be proved, provided the initial data are "small". The precise meaning of "small initial data" is given in the following definition.

Definition 2.2. The initial data on $\mathcal{C} = C_0 \cup \underline{C}_0$ are "small" if the following quantity, $J^{(q)}_{C_0 \cup \underline{C}_0}$, is small:

$$J^{(q)}_{C_0 \cup \underline{C}_0} = J^{(q)}_{C_0}\left[\overline{\gamma}_{ab}, \overline{\Omega}, \overline{\zeta}_a, \underline{X}, \underline{\omega}\right] + J^{(q)}_{\underline{C}_0}\left[\overline{\gamma}_{ab}, \underline{\Omega}, \underline{X}^a, \overline{\zeta}_a, \chi, \omega\right] \leq \epsilon \, , \tag{11}$$

where, with $p \in [2, 4]$ and $\delta > 0$,

$$J^{(q)}_{C_0}\left[\overline{\gamma}_{ab}, \overline{\Omega}, \overline{\zeta}_a, \underline{X}, \underline{\omega}\right] = \sup_{C_0}\left(r|\overline{\Omega} - \frac{1}{2}| + \frac{r^2}{\log r}|\mathrm{tr}\underline{X} - \frac{2}{r}| + r^{(\frac{5}{2}+\delta)}|\underline{\hat{\chi}}|\right) +$$

$$\sup_{C_0}\left[\left(\sum_{l=1}^{q}|r^{(2+l-\frac{2}{p})}\nabla^l\mathrm{tr}\chi|_{p,S} + \sum_{l=1}^{q}|r^{(\frac{5}{2}+l+\delta-\frac{2}{p})}\nabla^l\hat{\chi}|_{p,S} + \sum_{l=0}^{q}|r^{(\frac{7}{2}+l+\delta-\frac{2}{p})}\nabla^l\mathbf{D}_4\hat{\chi}|_{p,S}\right.$$

$$|r^{(2-\frac{2}{p})}\underline{\omega}|_{p,S} + \sum_{l=1}^{q}|r^{(2+l+\delta-\frac{2}{p})}\nabla^l\underline{\omega}|_{p,S}\bigg) + \sum_{l=1}^{q-1}|r^{(1+l+\delta-\frac{2}{p})}\nabla^l\log\overline{\Omega}|_{p,S}$$

$$+ \sum_{l=0}^{q-1}|r^{(2+l-\frac{2}{p})}\nabla^l\overline{\zeta}|_{p,S} + \sum_{l=0}^{q-2}|r^{(2+l-\frac{2}{p})}\nabla^l\underline{\omega}|_{p,S} + \sum_{l=0}^{q-2}|r^{(1+l-\frac{2}{p})}\nabla^l\underline{X}|_{p,S}\bigg] \, .$$

A similar expression holds for $J_{\underline{C}_0}^{(q)}\left[\overline{\gamma}_{ab}, \overline{\underline{\Omega}}, \overline{X}^a, \overline{\zeta}_a, \overline{\chi}, \overline{\omega}\right]$, see.[Ca-Ni2]

The $|\cdot|_{p,S}$ norms are the usual Sobolev norms on the two dimensional surfaces $S_0(\nu)$ and $\underline{S}_0(\lambda)$. The value of q is tied to the regularity of our solution. In the present proof it has to be greater than a fixed value, ≥ 7.

Due to the constraint equations for the initial data, to satisfy the smallness condition 11 it will be enough to impose that the norms of the quantities which are assigned freely on C_0 and $\underline{C}_0$ be small, together with the smallnesss of some norms relative to $S_0 = \underline{S}_0 = C_0 \cap \underline{C}_0$. These restricted conditions plus the transport equations 7, 8 along C_0 and $\underline{C}_0$, respectively, will imply the smallness of 11. Once the smallness condition for the initial data is defined we can state the characteristic global existence theorem.

Theorem 2.1. *Let the initial data*

$$\left\{C_0 \;; \overline{\gamma}_{ab}, \overline{\Omega}, \overline{\zeta}_a, \underline{X}_{ab}, \overline{\omega}\right\} \cup \left\{\underline{C}_0 \;; \overline{\gamma}_{ab}, \overline{\underline{\Omega}}, \overline{X}^a \overline{\zeta}_a, \overline{X}_{ab}, \overline{\omega}\right\} \tag{12}$$

be assigned together with their partial tangential derivatives to a fixed order specified by the integer $q \geq 7$. Let us assume they satisfy the smallness conditions:

$$J_{C_0 \cup \underline{C}_0}^{(q)} = J_{C_0}^{(q)}\left[\overline{\gamma}_{ab}, \overline{\Omega}, \overline{\zeta}^a, \underline{X}, \overline{\omega}\right] + J_{\underline{C}_0}^{(q)}\left[\overline{\gamma}_{ab}, \overline{\underline{\Omega}}, \overline{\zeta}_a, \overline{X}, \overline{\omega}\right] \leq \epsilon . \tag{13}$$

Then there exists and it is unique a vacuum Einstein spacetime $\{\mathcal{M}, \mathbf{g}\}$ solving the characteristic initial value problem with initial data 12. $\mathcal{M}$ is the maximal future development of $C(\lambda_1) \cup \underline{C}(\nu_0)$ [*],

$$\mathcal{M} = \lim_{\nu_1 \to \infty} J^+(S(\lambda_1, \nu_0)) \cap J^-(S(\lambda_0, \nu_1)) \equiv \lim_{\nu_1 \to \infty} \mathcal{M}(\nu_1) . \tag{14}$$

Moreover $\mathcal{M}$ is endowed with the following structures:

a) $\mathcal{M}(\nu_1)$ is foliated by a "double null canonical foliation" $\{C(\lambda)\}$, $\{\underline{C}(\nu)\}$, with $\lambda \in [\lambda_1, \lambda_0]$, $\nu \in [\nu_0, \nu_1]$ [*]. Double canonical foliation means that the null hypersurfaces $C(\lambda)$ are the level hypersurfaces of a function $u(p)$ solution of the eikonal equation

$$g^{\mu\nu}\partial_\mu w \partial_\nu w = 0 ,$$

with initial data a function $u_*(p)$ defining the foliation of the "final" incoming cone $\underline{C}(\nu_1)$, while the the null hypersurfaces $\underline{C}(\nu)$ are the level hypersurfaces of a function $\underline{u}(p)$ solution of the eikonal equation with initial data a function $\underline{u}_{(0)}(p)$ defining a canonical foliation of the initial outgoing cone $C(\lambda_1)$ [†]. The family of

---

[*]$C(\lambda_1)$ and $\underline{C}(\nu_0)$ are the initial cones C_0 and $\underline{C}_0$ thought imbedded in $\{\mathcal{M}, \mathbf{g}\}$, analogously $S_0 = S(\lambda_1, \nu_0) = C(\lambda_1) \cap \underline{C}(\nu_0)$.

[*]To avoid any misunderstanding "double canonical foliation" refers to a foliation of the spacetime $(\mathcal{M}, \mathbf{g})$, while with canonical foliation we denote a specific foliation of the initial data on C_0. Of course the first is related to the second.

[†]The "canonical" foliation on the portion $C(\lambda_1)$ of the initial hypersurface is not the one given when the initial data are specified, but it has been proved elsewhere that given the initial data foliation of $C(\lambda_1)$, it is possible to build on $C(\lambda_1)$ a "canonical" foliation. Its precise definition and the way for doing it is in.[Ni]

two dimensional surfaces $\{S(\lambda,\nu)\}$, where $S(\lambda,\nu) \equiv C(\lambda) \cap \underline{C}(\nu)$, defines a two dimensional foliation of $\mathcal{M}$.

b) $i(C_0) = C(\lambda_1)$, $i(\underline{C}_0) = \underline{C}(\nu_0)$, $i(S_0(\nu_0)) = i(\underline{S}_0(\lambda_1)) = C(\lambda_1) \cap \underline{C}(\nu_0)$.

c) On $C(\lambda_1)$ with respect to the initial data foliation[‡], we have

$$i^*(\gamma) = \overline{\gamma} \ , \ i^*(\Omega) = \overline{\Omega} \ , \ i_*^{-1}(X) = \overline{X}$$
$$i^*(\chi) = \overline{\chi} \ , \ i^*(\omega) = \overline{\omega} \ , \ i^*(\zeta) = \overline{\zeta} \tag{15}$$

together with their tangential derivatives up to q.

d) On $\underline{C}(\nu_0)$ with respect to the initial data foliation, we have

$$i^*(\gamma) = \overline{\gamma} \ , \ i^*(\Omega) = \underline{\overline{\Omega}}, \ i_*^{-1}(\underline{X}) = \overline{\underline{X}}$$
$$i^*(\underline{\chi}) = \underline{\overline{\chi}} \ , \ i^*(\underline{\omega}) = \underline{\overline{\omega}} \ , \ i^*(\zeta) = \overline{\zeta} \tag{16}$$

where $\gamma, \Omega, X, \chi, \underline{\chi}, \omega, \underline{\omega}, \zeta$ are the metric components and the connection coefficients in a neighbourhood of $C(\lambda_1)$ and $\underline{C}(\nu_0)$.

e) The constraint equations 7, 8 are the pull back of (some of) the structure equations of $\mathcal{M}$ restricted to $C(\lambda_1)$ and $\underline{C}(\nu_0)$[§].

f) The double null canonical foliation and the associated two dimensional one, $\{S(\lambda,\nu)\}$, implies different foliations of $C(\lambda_1)$ and $\underline{C}(\nu_0)$; with respect to this new foliation we can define on the whole $\mathcal{M}$ a null orthonormal frame $\{e_4, e_3, e_a\}$ adapted to it, the metric components, γ_{ab}, Ω, X_a with respect to the adapted coordinates $\{u, \underline{u}, \theta, \phi\}^*$ and the corresponding connection coefficients.

Moreover denoting N and $\underline{N}$ two null vector fields equivariant with respects to the $S(\lambda,\nu)$ surfaces, leaves of the canonical foliation, the following relations hold

$$N = 2\Omega^2 \frac{\partial}{\partial \underline{u}} \ , \ \underline{N} = 2\Omega^2 \left(\frac{\partial}{\partial u} + X^a \frac{\partial}{\partial w^a} \right)$$

with $X^a = g^{ab} X_b$ a vector field defined in $\mathcal{M}$. g) Finally the null geodesics of $\mathcal{M}$ along the outgoing and incoming null direction e_4, e_3 are defined for all $\nu \in [\nu_0, \infty)$ and all $\lambda \in [\lambda_1, \lambda_0]$ respectively.

Again we do not discuss here this result and we refer for its detailed and long proof to.[Ca-Ni2] We just recall that it is an adaptation of the global existence result proved in.[Kl-Ni2]

[‡]Observe that while $\mathcal{M}$ is globally foliated by a double null canonical foliation, it is not possible to foliate it with a double foliation solution of the eikonal equation with, as initial data the Ω-foliation of C_0 and $\underline{C}_0$. This is possible only in a small neighbourhood of C_0 and $\underline{C}_0$.

[§]With respect to the foliations of the initial data.

[*]For the precise way the coordinate θ and ϕ are defined see,[Kl-Ni2] paragraph 3.1.6.

166

2.3. *The local existence problem*

As said before some local existence results for this characteristic problem exist in the literature. They have the common property of being obtained in the "harmonic gauge" transforming the problem in a characteristic quasilinear hyperbolic one, generalization of a non linear scalar wave equation. This strategy has the drawback of not exploiting completely the geometric structure of the Einstein equations and, moreover, makes going from a local solution to a global one a difficult task[†]. Our result (in progress) follows the same path used to prove the constraint equations and the global existence, namely does not use the harmonic gauge and is based on the construction of a double foliation made by null outgoing and incoming cones. Without presenting any detail of the proof we try to give a thorough description of the main steps required for obtaining this result. They are the following:

i) Starting from the Einstein equations written as transport equations for the connection coefficients we have to transform them into a first order system of equations. This requires to use coordinates adapted to the foliation and to add to these equations those which connect the connection coefficients to the metric components γ, Ω, X. Once this is done we can look for a Cauchy-Kowalevski solution for this system.

ii) This requires to adapt the Cauchy-Kowalewski theorem to the characteristic case. This result has been obtained by Duff,,[Du] for a linear system and extended to the non linear situation by H.Friedrich,[Fr1] Moreover to use this result we need also to prove that the initial data satisfying the constraint equations and belonging to some Sobolev spacetime can be approximated by a sequence of real analytic initial data satisfying the same constraints equations.

iii) Although the initial claim was to provide a method basically coordinate independent it is clear that once we have to obtain a local solution some coordinates have to be used. Here we use the coordinate adapted to the double foliation which, in some sense, can be thought as our gauge choice. Also in this case, as it happens for the reduced Einstein equations in the harmonic gauge, one has to prove that the real analytic solution we obtain is a solution of the Einstein equations provided the initial data satisfy the constraints and a "gauge condition"[*].

iv) Once the previous steps have been satisfied we have a sequence $\{\Psi_n^{(0)}\}$ of real analytic initial data producing a sequence $\{\Psi_n\}$, of real analytic solution to our problem. To show that they can approximate a Sobolev solution we have to show first that all that these solution have a common existence domain. This fact is not provided by the Cauchy-Kowalevski theorem, but by the hyperbolicity of the problem we are considering. In the case of the non linear wave equation, for instance, one can first linearize the problem, obtain a real analytic sequence of so-

---

[†]Nevertheless it has been accomplished in the non characteristic case by H.Lindblad and I.Rodnianski,[Ro-Li]

[*]The analogous of the condition $\Gamma^\alpha = 0$ which in the harmonic case has to be proved satisfied in the whole spacetime if it is satisfied by the initial data.

lutions of the linearized problem, prove, easily, that there is a common existence region for all these solution and finally prove that their limit in an appropriate Sobolev norm is the required (Sobolev) solution. To obtain this result the hyperbolicity of the problem is an essential ingredient.

v) To extend this procedure from the non linear wave equation to the Einstein equations is the more delicate technical task. The reason is that the hyperbolicity of the Einstein equations written in an arbitrary gauge is in some sense more hidden. Therefore some work has to be done to find the norms which in the Einstein case play the role of the energy-type norms and which are bounded in terms of the initial data. In fact as discussed at length in$^{\text{Ch-Kl}}$ and$^{\text{Kl-Ni2}}$ these norms are written in terms of the Riemann components and require the use of the Bianchi equations to prove that they are bounded. Moreover the linearization procedure cannot be used as in that case the linearized solutions are not solutions of the vacuum Einstein equations* and , therefore, we cannot use their hyperbolicity and prove that the real analytic solutions have a common existence region.

vi) The problem of the common existence region is, therefore, faced directly for the non linear problem. This requires some technical work, but, in some sense, it is simply an extension of a result S.Klainerman and the present author already proved some years ago for the Burger equation, see,$^{\text{Kl-Ni1}}$ where the main ideas about the way to extend this result to the Einstein equations are discussed. Proving this result requires to use some results obtained in,$^{\text{Kl-Ni2}}$ namely one has to prove that the $|\cdot|_{p,S}$ Sobolev norms for the tangential derivatives of the connection coefficients of all order have to satisfy some appropriate estimates which allow to enlarge the analiticity region of the Cauchy-Kowalevski solutions. This result can be thought as the analog of some results obtained by S.Alinhac, G.Metivier,,$^{\text{A-M}}$ in the case of the Einstein equations.

vii) The previous step is one of the crucial step to prove our local existence result. Going back to the Burger equation one makes a proof by induction. Assume that some L^2 norms for the tangential (to a time constant hypersurface) derivatives up to order $N-1$ of the solution satisfy some detailed bounds for any time in the existence interval $[0, T']$ of the real analytic solution and prove by induction that this is true also for the order N derivatives. This allows to have a complete control of the solution for $t = T'$ and show that even for $t = T'$ the solution is real analytic and, again by Cauchy-Kowalevski, can be further extended up to a time T. To achieve the inductive proof the Burger equation is used to have, via Gronwall's Lemma, the control of the L^2 norms together with the fact that due to the hyperbolicity of the equation the first tangential derivatives satisfy some a priori estimate up to the time T which depends only on the Sobolev norm of the initial data.

viii) To extend this result to the Einstein equations requires, therefore, the use of their hyperbolicity through the energy-type norms made with the Bel-Robinson tensor and the use of the appropriate transport equations which allow to develop

---

*The analogous of the condition $\Gamma^\alpha = 0$ cannot be maintained in the linearized version of our equations.

the proof by induction. In broad terms we have to use those structure equations that do not imply a "loss of derivatives". This is somewhat delicate, but on the other side is just an extension (with some simplifications associated to the fact that looking for a local solution we do not have to worry about the decays of the solution along different directions) of what has been done in,[Kl-Ni2] Chapter 4.

ix) Once the previous step is achieved one has to show that using the previous Sobolev norms with a well defined s one can prove that the sequence of real analytic solutions is in a common ball of the Sobolev space and moreover it is a Cauchy sequence *. Then it is a standard procedure to show that the limit function is a solution of the same equations.

x) Finally this approach can be also used (in a very simplified case) to prove that the initial (Sobolev) data can be obtained as limit of a sequence of real analytic data.

References

A-M. S.Alinhac, G.Metivier, *"Propagation de l'analyticité des solutions de systèmes hyperboliques non-linéaires"*. Invent. Math. 75, (1984), 189-204.

Br1. Y.Choquet-Bruhat, *"Théorème d' existence pour certain systèmes d'équations aux dérivcés partielles nonlinéaires"*. Acta Matematica 88, (1952), 141-225.

Br-Ch2. Y.Choquet-Bruhat, D.Christodoulou, *"Elliptic systems in $H_{s,\delta}$ spaces on manifolds which are euclidean at infinity"*. Acta Math. 145, (1981), 129-150.

Ca-Ni1. G.Caciotta, F.Nicolò *"Global characteristic problem for Einstein vacuum equations with small initial data: (I) the Initial data constraints"*. JHDE 2, n.1 (2005), 201-277.

Ca-Ni2. G.Caciotta, F.Nicolò: *"Global characteristic problem for the Einstein vacuum equations with small initial data, (II): The existence proof"*. gr-qc/0608038.

Ch2. D.Christodoulou, *"The formation of black holes and singularities in spherically symmetric gravitational collapse"*. Comm.Pure appl.Math.44, (1991), 339-373.

Ch-Kl. D.Christodoulou, S.Klainerman, *"The global non linear stability of the Minkowski space"*. Princeton Mathematical series, 41 (1993).

Ch-Mu. D.Christodoulou, H.Muller Zum Hagen, *"Problème de valeur initiale caractéristique pour des systemes quasi linèaires du second ordre"*. C.R. Acad. Sc. Paris, t. 293 (1981), 39-42.

Do. M.Dossa, *"Problèmes de Cauchy sur un conoide caractéristique pour les equations d'Einstein (conformes) du vide et pour les equations de Yang-Mills-Higgs"*. Ann.Inst. H.Poincare' 4, (2003), 385-411.

Du. G.F.D.Duff, *"Mixed problems for linear systems of first order equations"*. Canadian Journal of Mathématics Vol. X, (1958), 127-160.

Fr1. H.Friedrich, *"On the existence of analytic null asymptotically flat solutions of Einstein's vacuum field equations"*. Proc.Roy.Soc.Lond. A381, (1982), 361-371.

Haw-El. S.W.Hawking, G.F.R.Hellis "The Large Scale Structure of Space-time" *Cambridge Monographs on Mathematical Physics*, 1973.

Kl-Ni1. S.Klainerman, F.Nicolò "On local and global aspects of the Cauchy problem in General Relativity" Classical Quantum Gravity16,(1999), R73-R157.

Kl-Ni2. S.Klainerman, F.Nicolò, *"The evolution problem in General Relativity"*. Birkhauser, Progress in Matehematical Physics, Vol.25 2002.

Mu-Se. H.Muller Zum Hagen, H.J.Seifert *"On characteristic initial value and mixed problems"*. Gen.Rel. and Grav. 8, n.4, (1977), 259-301.

Mu. H.Muller Zum Hagen, *"Characteristic initial value problem for hyperbolic systems of second order differential equations"*. Ann.Inst. H.Poincare' 53, n.2, (1990), 159-216.

Ni. F.Nicolò *"Canonical foliation on a null hypersurface"*. Journal of hyperbolic differential equations Vol.1, n.3 (2004) 1-62.

Ren.   A.D.Rendall, *"Reduction of the characteristic initial value problem to the Cauchy problem and its applications to the Einstein equations"*. Proc.Roy.Soc.Lond. A427, (1990), 221-239.

Ro-Li.   H.Lindblad, I.Rodnianski *"Global existence for the Einstein vacuum equations in wave coordinates"*. math.AP/0312479, submitted to CMP (2004)

Sp.   M.Spivak, *"A comprehensive introduction to Differential Geometry"*. Wilmington: Publish or Perish, Inc. (1970).

LONG TIME BEHAVIOUR OF THREE COMPETING SPECIES AND MUTUALISTIC COMMUNITIES

Salvatore RIONERO

University of Naples Federico II
Department of Mathematics and Applications "R. Caccioppoli"
Complesso Universitario Monte S. Angelo - Via Cinzia, 80126 Naples - ITALY
E-mail:rionero@unina.it

1. Introduction

Let $\Omega \subset I\!\!R^3$ be a smooth bounded habitat of a population S constituted by three species S_i ($i = 1, 2, 3$). The spazio-temporal dynamics of S is very often modelled by a dimensionless reaction-diffusion system like {Cfr. [1]-[5] and references therein}

$$\frac{\partial U_i}{\partial t} = \sum_{j=1}^{3} a_{ij} U_j + d_i \Delta U_i + N_i(U_1, U_2, U_3) \tag{1}$$

with

$$U_i \; : \; (\mathbf{x}, t) \in \Omega \times I\!\!R^+ \Rightarrow U_i(\mathbf{x}, t) \in I\!\!R^+ \qquad i = 1, 2, 3$$

density of S_i and

- $a_{ij}, d_i > 0$ constants, $i, j \in \{1, 2, 3\}$;

- N_i ($i = 1, 2, 3$) nonlinear functions such that the system

$$\sum_{j=1}^{3} a_{ij} U_j + N_i(U_1, U_2, U_3) = 0 \tag{2}$$

admits (at least) a biologically meaningful solution $\overline{\mathbf{U}} = (\overline{U}_1, \overline{U}_2, \overline{U}_3)$ with $\overline{U}_i \geq 0$ ($i = 1, 2, 3$) i.e. an ecological equilibrium state. Then $\overline{\mathbf{U}}$ is a critical point of (1) under the Robin boundary data

$$U_i + \left(\frac{1}{\beta_i} - 1\right) \nabla U_i \cdot \mathbf{n} = \overline{U}_i, \qquad \text{on } \Sigma \tag{3}$$

with $\beta_i =$const.$\in]0, 1]$, $\Sigma = \partial \Omega$ and $\mathbf{n}$ the unit outward normal to Σ.

Setting

$$U_i = \overline{U}_i + u_i, \qquad i = 1, 2, 3 \tag{4}$$

the long time behaviour of S depends on the long time behaviour of the solutions to the initial boundary value problem

$$\frac{\partial u_i}{\partial t} = \sum_{j=1}^{3} a_{ij} u_i + d_i \Delta u_i + N_i^*(u_1, u_2, u_3) \tag{5}$$

$$u_i + \left(\frac{1}{\beta_i} - 1\right) \nabla u_i \cdot \mathbf{n} = 0 \qquad \text{on } \Sigma \tag{6}$$

with $N_i^* = N_i(\mathbf{U}) - N_i(\overline{\mathbf{U}})$.

We denote by

- $\langle \cdot, \cdot \rangle$ the scalar product in $L^2(\Omega)$;
- $\langle \cdot, \cdot \rangle_\Sigma$ the scalar product in $L^2(\Sigma)$
- $\| \cdot \|$ the $L^2(\Omega)$-norm
- $\| \cdot \|_\Sigma$ the $L^2(\Sigma)$-norm
- $W_i^{1,2}(\Omega, \Sigma, \beta_i)$ the functional space such that

$$\varphi \in W_i^{1,2}(\Omega, \Sigma, \beta_i) \to \{\varphi \in W_i^{1,2}(\Omega) \cap W_i^{1,2}(\Sigma), (6) \text{ hold}\} \tag{7}$$

- $\overline{\alpha}_i$ the positive constant appearing in the inequality $(i = 1, 2)$ [4], [6]

$$\|\nabla \varphi\|^2 + \frac{\beta_i}{1 - \beta_i} \|\varphi\|_\Sigma^2 \geq \overline{\alpha}_i \|\varphi\|^2, \tag{8}$$

holding in $W_i^{1,2}(\Omega, \Sigma, \beta_i)$.

As it is well known $\overline{\alpha}_i(\Omega, \beta_i)$ is the lowest eigenvalue of

$$\Delta \varphi + \lambda \varphi = 0 \tag{9}$$

in $W_i^{1,2}(\Omega, \Sigma, \beta_i)$ (i.e. the principal eigenvalue of $-\Delta$).

Our aim is to characterize the solutions long time behaviour by giving conditions necessary and sufficient for the $L^2(\Omega)$-stability of $\overline{\mathbf{U}}$. Precisely – setting

$$\begin{cases} b_{ii} = a_{ii} - d_i \overline{\alpha}_i, & (i = 1, 2, 3) \\ b_{ij} = a_{ij}, & (i \neq j = 1, 2, 3) \end{cases} \tag{10}$$

our aim is to show that, under the assumptions

$$a_{ij} a_{ji} > 0 \qquad i \neq j \tag{11}$$

the following theorems hold.

Theorem 1 - *Let (11) and*

$$\begin{cases} a_{12} a_{23} a_{31} = a_{13} a_{21} a_{32} \\ \\ \mathbf{N}^*(\mathbf{u}) = o(\|\mathbf{u}\|) \end{cases} \tag{12}$$

with $\mathbf{u} = (u_1, u_2, u_3)$, $\mathbf{N}^* = (N_1^*, N_2^*, N_3^*)$ *hold. Then, with respect to the* $L^2(\Omega)$-*norm,* $\mathbf{U}$ *is stable if and only if is stable the zero solution of the linear system of O.D.Es.*

$$\frac{d\xi_i}{dt} = \sum_{j=1}^{3} b_{ij}\xi_j \qquad (i = 1, 2, 3) \tag{13}$$

Theorem 2 - *Let (11)-(12) hold. Then the zero solution of (13) is stable if and only if*

$$(-1)^k D_k > 0 \quad (k = 1, ..., n) \tag{14}$$

with D_k *principal minor of order* k *of the matrix* $\|b_{ij}\|$.

For the sake of simplicity we refer here only to the linear stability of the three species competing models. The nonlinear stability can be obtained by repeating the procedures used for the binary systems of P.D.Es.[8]-[10].

The plan of the paper is as follows. Section 2 is devoted to show that, by suitable scalings, the matrix $\|b_{ij}\|$ is symmetrizable. Further the Jacobi's reduction to the canonical form is recalled. In Section 3 stability-instability is considered for an associate system of O.D.Es.. After the introduction of a three species competitive model (Section 4), in Section 5 we symmetrize the system and give the proof of Theorem 2. In Section 6 the absence of Turing instability for the competitive and mutualistic system is remarked. The paper ends (Section 7) with some final remarks concerning systems to which the obtained results can be applied or generalized.

2. A symmetrization theorem and Jacobi method of reducing a quadratic form to a sum of squares

Theorem 3 - *Let (11) hold. Then the eigenvalues of the matrix* $\|b_{ij}\|$ *are all real and coincide with the eigenvalues of the matrix* $\|\sigma_{ij}\|$ *with*

$$\begin{cases} \sigma_{ii} = b_{ii} & i = 1, 2, 3 \\[2ex] \sigma_{ij} = \pm\sqrt{a_{ij}a_{ji}} & i \neq j \end{cases} \tag{15}$$

Proof. By virtue of (15) it follows that

$$\begin{vmatrix} b_{11} & b_{12} \\ b_{21} & b_{22} \end{vmatrix} = b_{11}b_{22} - b_{12}b_{21} = \sigma_{11}\sigma_{22} - \sigma_{12}\sigma_{21}.$$

Setting

$$\frac{\delta_j}{\delta_i} = \sqrt{\frac{a_{ij}}{a_{ji}}} \qquad i \neq j$$

it turns out that

$$\begin{vmatrix} \sigma_{11} & \sigma_{12} & \sigma_{13} \\ \sigma_{21} & \sigma_{22} & \sigma_{23} \\ \sigma_{31} & \sigma_{32} & \sigma_{33} \end{vmatrix} = \frac{1}{\delta_1\delta_2\delta_3} \begin{vmatrix} \delta_1 b_{11} & \delta_2 a_{12} & \delta_3 a_{13} \\ \delta_1 a_{21} & \delta_2 b_{22} & \delta_3 a_{23} \\ \delta_1 a_{31} & \delta_2 a_{32} & \delta_3 b_{33} \end{vmatrix} = \begin{vmatrix} b_{11} & a_{12} & a_{13} \\ a_{21} & b_{22} & a_{23} \\ a_{31} & a_{32} & b_{33} \end{vmatrix}. \tag{16}$$

Then the matrices $\|b_{ij}\|$, $\|\sigma_{ij}\|$ have the same secular equation and hence the same eigenvalues. By virtue of the symmetry of $\|\sigma_{ij}\|$, all the eigenvalues are real.

Setting

$$A_i = \sum_{j=1}^{3} \sigma_{ij}\xi_j, \qquad (i = 1, 2, 3) \tag{17}$$

and introducing the Jacobi's variables X_i

$$\begin{cases} X_1 = A_1 \\[2mm] X_2 = \begin{vmatrix} \sigma_{11} & A_1 \\ \sigma_{21} & A_2 \end{vmatrix} \\[4mm] X_3 = \begin{vmatrix} \sigma_{11} & \sigma_{12} & A_1 \\ \sigma_{21} & \sigma_{22} & A_2 \\ \sigma_{31} & \sigma_{32} & A_3 \end{vmatrix} \end{cases} \tag{18}$$

the symmetric quadratic form

$$Q = \sum_{i,j}^{1-3} \sigma_{ij}\xi_i\xi_j \tag{19}$$

becomes [7]

$$Q = \sum_{k=1}^{3} \frac{X_k}{D_{k-1}D_k} = \frac{X_1^2}{\sigma_{11}} + \frac{X_2^2}{\sigma_{11}D_2} + \frac{X_3^2}{D_2 D_3} \tag{20}$$

with

$$\begin{cases} D_0 = 1, \qquad\qquad\qquad D_1 = b_{11} = \sigma_{11} \\[4mm] D_2 = \begin{vmatrix} \sigma_{11} & \sigma_{12} \\ \sigma_{21} & \sigma_{22} \end{vmatrix}, \quad D_3 = \begin{vmatrix} \sigma_{11} & \sigma_{12} & \sigma_{13} \\ \sigma_{21} & \sigma_{22} & \sigma_{23} \\ \sigma_{31} & \sigma_{32} & \sigma_{33} \end{vmatrix} \end{cases} \tag{21}$$

and hence (20) is negative definite only if

$$\begin{cases} D_1 < 0 \\ D_2 > 0 \\ D_3 < 0. \end{cases} \tag{22}$$

We end by observing that the system (18) can be written

$$\begin{cases} \sigma_{11}\xi_1 + \sigma_{12}\xi_2 + \sigma_{13}\xi_3 = X_1 \\ \gamma_{22}\xi_2 + \gamma_{23}\xi_3 = X_2 \\ \gamma_{33}\xi_3 = X_3 \end{cases} \tag{23}$$

with

$$\begin{cases} \gamma_{22} = D_2 \\ \gamma_{23} = \sigma_{13}\sigma_{21} - \sigma_{11}\sigma_{23} \\ \gamma_{33} = D_3\,. \end{cases} \tag{24}$$

Then (22) imply that (23) is uniquely solvable and it turns out that

$$\begin{cases} \xi_1 = \dfrac{1}{\sigma_{11}}\left[X_1 - \dfrac{\sigma_{12}}{D_2}X_2 + \dfrac{1}{D_3}(\sigma_{12}\gamma_{23} - \sigma_{13})X_3\right] \\[2mm] \xi_2 = \dfrac{1}{D_2}(X_2 - \dfrac{\gamma_{23}}{D_3}X_3) \\[2mm] \xi_3 = \dfrac{1}{D_3}X_3\,. \end{cases} \tag{25}$$

Remark 1 - *In view of the fact that the terms of (19) can be ordered in different manner, (19) is definite negative if all the principal minor of order 1 are negatives, all the principal minor of order 2 are positive and $\det\|\sigma_{ij}\| < 0$. Then immediately follows that $\sigma_{11} + \sigma_{22} + \sigma_{33} < 0$.*

Remark 2 - *We observe that $\dfrac{\delta_j}{\delta_i} = \sqrt{\dfrac{a_{ij}}{a_{ji}}}$ and $(12)_1$ guarantee the consistence of*

$$\frac{\delta_1}{\delta_3} = \frac{\delta_1}{\delta_2} \cdot \frac{\delta_2}{\delta_3}\,.$$

Remark 3 - *(3) can be written*

$$\beta_i U_i + (1 - \beta_i)\nabla U_i \cdot \mathbf{n} = \beta_i \overline{U}_i\,.$$

Therefore the excluded case $\beta_i = 0$, implies

$$\nabla U_i \cdot \mathbf{n} = 0$$

176

i.e. the Neumann homogeneous boundary conditions on U_i. This case, already considered in [9], [11], can be included also here, but, in order to avoid the eigenvalues $\overline{\alpha}_i = 0$ $(i = 1, 2, 3)$ either $(\overline{U}_1, \overline{U}_2, \overline{U}_3)$ has to be the unique equilibrium solution or only perturbations with zero mean value on Ω can be considered.

Remark 4 - *The case $a_{ij}a_{ji} = 0$ for some (i, j) is studied in the last section.*

3. Stability - Instability conditions

Theorem 4 - *Let (22) hold. Then*

$$E = \frac{1}{2}(\xi_1^2 + \xi_2^2 + \xi_3^2) \tag{26}$$

with ξ_i $(i = 1, 2, 3)$ given by (25), is a Liapunov function for the system

$$\frac{d\xi_i}{dt} = \sum_{j=1}^{3} \sigma_{ij}\xi_j \qquad (i = 1, 2, 3) \tag{27}$$

and – along the solutions of (27) – it turns out that

$$E \leq E_0 e^{-dt} \tag{28}$$

with

$$0 < d \leq 3(\sigma_{11}^2 + \sigma_{12}^2 + \sigma_{13}^2 + D_2^2 + D_3^2)\inf\left(\frac{1}{|\sigma_{11}|}, \frac{1}{|\sigma_{11}D_2|}, \frac{1}{|D_2D_3|}\right) \tag{29}$$

Proof. By virtue of (27)

$$\frac{1}{2}\frac{d}{dt}(\xi_1^2 + \xi_2^2 + \xi_3^2) = \sum_{i,j}^{1-3} \sigma_{ij}\xi_i\xi_j \tag{30}$$

hence (20) implies

$$\frac{dE}{dt} = \frac{X_1^2}{D_1} + \frac{X_2^2}{D_1 D_2} + \frac{X_3^2}{D_2 D_3} . \tag{31}$$

In view of

$$X_i^2 \leq mE \tag{32}$$

with

$$m = 3(\sigma_{11}^2 + \sigma_{12}^2 + \sigma_{13}^2 + D_2^2 + D_3^2) \tag{33}$$

from (31), (28) immediately follows.

Theorem 5 - *If one of the inequalities (22) is not verified, then the zero solution of (27) is unstable.*

Proof. In view of (31), the instability is an immediate consequence of the Liapunov-Chetayev instability theorem.

4. Long time behaviour of a three species competition model

Let us begin by considering the general three species competition model

$$
\begin{cases}
\dfrac{\partial U_1}{\partial t} = d_1 \Delta U_1 + U_1(a_1 - U_1 - \delta_{12}U_2 - \delta_{13}U_3) - f_1(U_1, U_2, U_3) \\[2mm]
\dfrac{\partial U_2}{\partial t} = d_2 \Delta U_2 + U_2(a_2 - \delta_{21}U_1 - U_2 - \delta_{23}U_3) - f_2(U_1, U_2, U_3) \\[2mm]
\dfrac{\partial U_3}{\partial t} = d_3 \Delta U_3 + U_3(a_3 - \delta_{31}U_1 - \delta_{32}U_2 - U_3) - f_3(U_1, U_2, U_3)
\end{cases}
\tag{34}
$$

with $f_i \geq 0$ for $U_i \geq 0 \ \forall i \in \{1,2,3\}$ and all the parameters d_i, a_i, δ_{ij} $(i, j = 1, 2, 3)$ positive constants.

The permanence of all the species S_i depends on the stability of the solution $(0, 0, 0)$ with respect to the perturbations $\mathbf{U} = (U_1, U_2, U_3)$ biologically meaningful, i.e. with $U_i \geq 0 \ (i = 1, 2, 3)$ verifing (34) and (3) with $\overline{U}_i = 0 \ (i = 1, 2, 3)$. It turns out that

$$
\begin{cases}
\dfrac{\partial U_1}{\partial t} \leq d_1 \Delta U_1 + a_1 U_1 \\[2mm]
\dfrac{\partial U_2}{\partial t} \leq d_2 \Delta U_2 + a_2 U_2 \\[2mm]
\dfrac{\partial U_3}{\partial t} \leq d_3 \Delta U_3 + a_3 U_3
\end{cases}
\tag{35}
$$

and hence, in view of (8), one obtains

$$
\begin{cases}
\dfrac{1}{2}\dfrac{d}{dt}\|U_1\|^2 \leq b_{11}\|U_1\|^2 \\[2mm]
\dfrac{1}{2}\dfrac{d}{dt}\|U_2\|^2 \leq b_{22}\|U_2\|^2 \\[2mm]
\dfrac{1}{2}\dfrac{d}{dt}\|U_3\|^2 \leq b_{33}\|U_3\|^2
\end{cases}
\tag{36}
$$

with (according to $(10)_1$)

$$
\begin{cases}
b_{11} = a_1 - d_1\overline{\alpha}_1 \\[2mm]
b_{22} = a_2 - d_2\overline{\alpha}_2 \\[2mm]
b_{33} = a_3 - d_3\overline{\alpha}_3.
\end{cases}
\tag{37}
$$

Therefore it follows that

i) $b_{ii} < 0$ implies the extinction of species S_i

ii) $b_{ii} < 0$, $\forall i \in \{1, 2, 3\}$, imply the three species extintion

iii) $b_{ii} > 0$ implies the instability and hence the permanence of S_i.

On passing to the case of the three species Lotka-Volterra competition model obtained by setting in (34)

$$f_i = 0 \qquad i = 1, 2, 3,$$ (38)

we assume that the parameters d_i, a_i, δ_{ij} verify $(12)_1$ and allow the existence of a positive solution $\overline{U} = (\overline{U}_1 > 0, \overline{U}_2 > 0, \overline{U}_3 > 0)$ of

$$\begin{cases} U_1 + \delta_{12}U_2 + \delta_{13}U_3 = a_1 \\ \delta_{21}U_1 + U_2 + \delta_{23}U_3 = a_2 \\ \delta_{31}U_1 + \delta_{32}U_2 + U_3 = a_3 \end{cases}$$ (39)

In view of (4), one linearizing obtains

$$\begin{cases} \dfrac{\partial u_1}{\partial t} = d_1 \Delta u_1 - \overline{U}_1(u_1 + \delta_{12}u_2 + \delta_{13}u_3) \\ \dfrac{\partial u_2}{\partial t} = d_2 \Delta u_2 - \overline{U}_2(\delta_{21}u_1 + u_2 + \delta_{23}u_3) \\ \dfrac{\partial u_3}{\partial t} = d_3 \Delta u_3 - \overline{U}_3(\delta_{31}u_1 + \delta_{32}u_2 + u_3) . \end{cases}$$ (40)

Introducing the scalings

$$\begin{cases} u_i = \alpha_i v_i \quad i, j = 1, 2, 3 \\ \alpha_i = const. \end{cases}$$ (41)

and setting

$$\frac{\alpha_i}{\alpha_j} = \mu_{ji} \qquad i, j = 1, 2, 3$$ (42)

it follows that

$$\begin{cases} \dfrac{\partial v_1}{\partial t} = d_1 \Delta v_1 - \overline{U}_1(v_1 + \delta_{12}\mu_{12}v_2 + \delta_{13}\mu_{13}v_3) \\ \dfrac{\partial v_2}{\partial t} = d_2 \Delta v_2 - \overline{U}_2(\delta_{21}\mu_{21}v_1 + v_2 + \delta_{23}\mu_{23}v_3) \\ \dfrac{\partial v_3}{\partial t} = d_3 \Delta v_3 - \overline{U}_3(\delta_{31}\mu_{31}v_1 + \delta_{32}\mu_{32}v_2 + v_3) . \end{cases}$$ (43)

Setting (according to (13))

$$\begin{cases} b_{11} = -(\overline{U}_1 + d_1\overline{\alpha}_1), \; b_{22} = -(\overline{U}_2 + d_2\overline{\alpha}_2), \; b_{33} = -(\overline{U}_3 + d_3\overline{\alpha}_3), \\ b_{12} = -\overline{U}_1\delta_{12}, \qquad b_{13} = -\overline{U}_1\delta_{13}, \\ b_{21} = -\overline{U}_2\delta_{21}, \qquad b_{23} = -\overline{U}_2\delta_{23}, \\ b_{31} = -\overline{U}_3\delta_{31}, \qquad b_{32} = -\overline{U}_3\delta_{32} \end{cases}$$ (44)

one obtains

$$\begin{cases} \dfrac{\partial v_1}{\partial t} = b_{11}v_1 + b_{12}\mu_{12}v_2 + b_{13}\mu_{13}v_3 + F_1 \\[2mm] \dfrac{\partial v_2}{\partial t} = b_{21}\mu_{21}v_1 + b_{22}v_2 + b_{23}\mu_{23}v_3 + F_2 \\[2mm] \dfrac{\partial v_3}{\partial t} = b_{31}\mu_{31}v_1 + b_{32}\mu_{32}v_2 + b_{33}v_3 + F_3 \end{cases} \tag{45}$$

with

$$\begin{cases} b_{ij}b_{ji} > 0 \qquad\qquad i = 1,2,3 \\[2mm] F_i = d_i(\Delta v_i + \overline{\alpha}_i v_i)\,. \end{cases} \tag{46}$$

5. Symmetrization and stability of $\overline{U}=(\overline{U}_1{>}0,\overline{U}_2{>}0,\overline{U}_3{>}0)$

In view of

$$\mu_{ji} = \frac{1}{\mu_{ij}} \tag{47}$$

choosing

$$\mu_{ij} = \sqrt{\frac{a_{ji}}{a_{ij}}} \tag{48}$$

it follows that

$$\frac{\partial}{\partial t}\begin{pmatrix} v_1 \\ v_2 \\ v_3 \end{pmatrix} = \begin{pmatrix} \sigma_{11} & \sigma_{12} & \sigma_{13} \\ \sigma_{21} & \sigma_{22} & \sigma_{23} \\ \sigma_{31} & \sigma_{32} & \sigma_{33} \end{pmatrix}\begin{pmatrix} v_1 \\ v_2 \\ v_3 \end{pmatrix} + \begin{pmatrix} F_1 \\ F_2 \\ F_3 \end{pmatrix} \tag{49}$$

with

$$\sigma_{ij} = -\sqrt{a_{ij}a_j}\,, \quad i \neq j \tag{50}$$

and hence, by virtue of (8), it turns out that

$$\frac{1}{2}\frac{d}{dt}(\|v_1\|^2 + \|v_2\|^2 + \|v_3\|^2) \leq \sum_{i,j}^{1-3} <\sigma_{ij}v_i, v_j>\,. \tag{51}$$

Proof. of Theorem 2. The proof of theorem 2 in the case at hand immediately follows. In fact let

$$\begin{cases} b_{11} < 0 \\ D_2 > 0 \\ D_3 < 0. \end{cases} \tag{52}$$

Then, in view of

$$\sum_{i,j}^{1-3}\sigma_{ij}v_iv_j = \frac{X_1^2}{\sigma_{11}} + \frac{X_2^2}{\sigma_{11}D_2} + \frac{X_3^2}{D_2 D_3} \tag{53}$$

with X_i Jacobi's variables, it follows that

$$\frac{1}{2}\frac{d}{dt}(\|v_1\|^2 + \|v_2\|^2 + \|v_3\|^2) \leq \frac{\|X_1\|^2}{\sigma_{11}} + \frac{\|X_2\|^2}{\sigma_{11}D_2} + \frac{\|X_3\|^2}{D_2 D_3} \tag{54}$$

and hence, with a procedure completely anologous to the which one followed in the proof of theorem 4, it follows that exists a positive constant d such that

$$E \le E_0 e^{-dt} \tag{55}$$

with

$$E = \frac{1}{2}\left(\|v_1\|^2 + \|v_2\|^2 + \|v_3\|^2\right). \tag{56}$$

6. Absence of Turing instability in competition and mutualistic systems

We refer here, for the sake of simplicity, to the case $n = 2$. Then the diffusion driven instability occurs iff

$$\begin{cases} I_0 = a_{11} + a_{22} < 0 \\[2mm] D_2^* = a_{11}a_{22} - a_{12}a_{21} > 0 \\[2mm] D_2 = b_{11}b_{22} - a_{12}a_{21} = d_1 d_2 \overline{\alpha}_1 \overline{\alpha}_2 - (d_1\overline{\alpha}_1 a_{22} + d_2\overline{\alpha}_2 a_{11}) + A_0 < 0. \end{cases} \tag{57}$$

The consistency of $(57)_3$ require

$$\begin{cases} d_1\overline{\alpha}_1 \ne d_2\overline{\alpha}_2 \\[2mm] a_{11}a_{22} < 0. \end{cases} \tag{58}$$

Therefore, in view of $(58)_2$ and (11), $(57)_2$ cannot occur and the absence of Turing instability also in the case $d_1\overline{\alpha}_1 \ne d_2\overline{\alpha}_2$ is immediately obtained.

7. Final remarks

We observe that

i) $(12)_1$ is verified in the remarkable cases

$$\begin{cases} \dfrac{\partial u_1}{\partial t} = a_{11}u_1 + d_1\Delta u_1 \\[3mm] \dfrac{\partial u_2}{\partial t} = a_{21}u_1 + a_{22}u_2 + a_{23}u_3 + d_2\Delta u_2 \\[3mm] \dfrac{\partial u_3}{\partial t} = a_{31}u_1 + a_{32}u_2 + a_{33}u_3 + d_3\Delta u_3 \end{cases} \tag{59}$$

$$\begin{cases} \dfrac{\partial u_1}{\partial t} = a_{11}u_1 + a_{12}u_2 + d_1\Delta u_1 \\[3mm] \dfrac{\partial u_2}{\partial t} = a_{21}u_1 + a_{22}u_2 + d_2\Delta u_2 \\[3mm] \dfrac{\partial u_3}{\partial t} = a_{31}u_1 + a_{32}u_2 + a_{33}u_3 + d_3\Delta u_3 \end{cases} \tag{60}$$

$$\begin{cases} \dfrac{\partial u_1}{\partial t} = a_{11}u_1 + a_{12}u_2 + d_1\Delta u_1 \\[2mm] \dfrac{\partial u_2}{\partial t} = a_{21}u_1 + a_{22}u_2 + a_{23}u_3 + d_2\Delta u_2 \\[2mm] \dfrac{\partial u_3}{\partial t} = a_{32}u_2 + a_{33}u_3 + d_3\Delta u_3 \,. \end{cases} \tag{61}$$

The systems (59), (60) can be split in two independent systems. As concerns (61) the symmetrization is obtained by choosing $\mu_{21} = \dfrac{\alpha_1}{\alpha_2} = \sqrt{\dfrac{b_{12}}{b_{21}}}$, $\mu_{32} = \dfrac{\alpha_2}{\alpha_3} = \sqrt{\dfrac{b_{13}}{b_{31}}}$;

ii) in the case $\{a_{13} = 0,\ a_{12}a_{23}a_{31} \neq 0\}$ following (10), (41), (42) and (46) we obtain (45) with $b_{13} = 0$, i.e.

$$\begin{cases} \dfrac{\partial v_1}{\partial t} = b_{11}v_1 + b_{12}\mu_{12}v_2 + F_1 \\[2mm] \dfrac{\partial v_2}{\partial t} = b_{21}\mu_{21}v_1 + b_{22}v_2 + b_{23}\mu_{23}v_3 + F_2 \\[2mm] \dfrac{\partial v_3}{\partial t} = b_{31}\mu_{31}v_1 + b_{32}\mu_{32}v_2 + b_{33}v_3 + F_3 \,. \end{cases} \tag{62}$$

Choosing

$$\begin{cases} \mu_{12} = \sqrt{\dfrac{b_{21}}{b_{12}}} = \dfrac{1}{\mu_{21}} \\[3mm] \mu_{23} = \sqrt{\dfrac{b_{32}}{b_{23}}} = \dfrac{1}{\mu_{32}} \\[3mm] \mu_{31} = \dfrac{\alpha_1}{\alpha_3} = \dfrac{\alpha_1}{\alpha_2} \cdot \dfrac{\alpha_2}{\alpha_3} = \sqrt{\dfrac{b_{12}}{b_{21}}} \cdot \sqrt{\dfrac{b_{23}}{b_{32}}} = \sqrt{\dfrac{b_{12}b_{23}}{b_{21}b_{32}}} \end{cases} \tag{63}$$

it turns out that

$$\begin{cases} \dfrac{\partial v_1}{\partial t} = b_{11}v_1 + \sqrt{b_{12}b_{21}}\,v_2 + F_1 \\[2mm] \dfrac{\partial v_2}{\partial t} = \sqrt{b_{12}b_{21}}\,v_1 + b_{22}v_2 + \sqrt{b_{23}b_{32}}\,v_3 + F_2 \\[2mm] \dfrac{\partial v_3}{\partial t} = 2b_{31}^{*}v_1 + \sqrt{b_{23}b_{32}}\,v_2 + b_{33}v_3 + F_3 \,, \end{cases} \tag{64}$$

with

$$b_{31}^{*} = \frac{1}{2}b_{31}\sqrt{\frac{b_{12}b_{32}}{b_{21}b_{32}}} \,. \tag{65}$$

By virtue of

$$\frac{1}{2}\frac{d}{dt}\left(\|v_1\|^2 + \|v_2\|^2 + \|v_3\|^2\right) \le b_{11}\|v_1\|^2 + b_{22}\|v_2\|^2 + b_{33}\|v_3\|^2$$
$$+2\sqrt{b_{12}b_{21}}\langle v_1, v_2\rangle + 2\sqrt{b_{23}b_{32}}\langle v_2, v_3\rangle + 2b_{31}^*\langle v_1, v_3\rangle. \tag{66}$$

The sign of the right hand side of (66) depends on the quadratic form

$$Q = \sum_1^3 p_{ij}\xi_i\xi_j \qquad I, J = 1, 2, 3 \tag{67}$$

with

$$p_{31} = p_{13} = b_{31}^* \tag{68}$$

and

$$p_{ij} = b_{ij} \tag{69}$$

in any other case and therefore on the principal minors of the matrix

$$\begin{pmatrix} p_{11} & p_{12} & p_{13} \\ p_{21} & p_{22} & p_{23} \\ p_{31} & p_{32} & p_{33} \end{pmatrix}. \tag{70}$$

Then the stability is guaranteed by

$$\begin{cases} \overline{D}_1 = p_{11} = b_{11} < 0 \\[2mm] \overline{D}_2 = p_{11}p_{22} - p_{12}p_{21} = b_{11}b_{22} - b_{12}b_{21} > 0 \\[2mm] \overline{D}_3 = \begin{vmatrix} p_{11} & p_{12} & p_{13} \\ p_{21} & p_{22} & p_{23} \\ p_{31} & p_{32} & p_{33} \end{vmatrix} < 0. \end{cases} \tag{71}$$

In order to obtain $\overline{D}_3 < 0$ directly in terms of the b_{ij} we observe that

$$\overline{D}_3 = \begin{vmatrix} b_{11} & \dfrac{\alpha_2}{\alpha_1}b_{12} & \dfrac{1}{2}\dfrac{\alpha_1}{\alpha_3}b_{31} \\[2mm] \dfrac{\alpha_1}{\alpha_2}b_{21} & b_{22} & \dfrac{\alpha_3}{\alpha_2}b_{23} \\[2mm] \dfrac{1}{2}\dfrac{\alpha_1}{\alpha_3}b_{31} & \dfrac{\alpha_2}{\alpha_3}b_{32} & b_{33} \end{vmatrix} =$$

$$= \dfrac{1}{\alpha_1\alpha_2\alpha_3} \begin{vmatrix} \alpha_1 b_{11} & \alpha_2 b_{12} & \dfrac{1}{2}\dfrac{\alpha_1^2}{\alpha_3}b_{31} \\[2mm] \alpha_1 b_{21} & \alpha_2 b_{22} & \alpha_3 b_{23} \\[2mm] \dfrac{1}{2}\alpha_1 b_{31} & \alpha_2 b_{32} & \alpha_3 b_{33} \end{vmatrix} = \tag{72}$$

$$= \begin{vmatrix} b_{11} & b_{12} & \dfrac{1}{2}\left(\dfrac{\alpha_1}{\alpha_3}\right)^2 b_{31} \\[2mm] b_{21} & b_{22} & b_{23} \\[2mm] \dfrac{1}{2}b_{31} & b_{32} & b_{33} \end{vmatrix}$$

with

$$\left(\frac{\alpha_1}{\alpha_3}\right)^2 = \frac{b_{12}b_{23}}{b_{21}b_{32}}; \tag{73}$$

iii) to the case ii) belongs the three component Orengonator model, for a light-sensitive Belousov-Zhabotinskii medium [5], [12]. In fact the equations governing the perturbations (u, v, w) to an equilibrium state $(\overline{u}, \overline{v}, \overline{w})$ with

$$\overline{u} = \overline{v} = \overline{w} = \frac{\phi + f\overline{u}}{\overline{u} + q} \tag{74}$$

are [13]

$$\begin{cases} w_t = (\varepsilon')^{-1}[-(q+\overline{u})w + fv - \overline{w}u + d_3\Delta w] \\[2mm] u_t = \varepsilon^{-1}[(q-\overline{u})w + (1-2\overline{u}-\overline{w})u + d_1\Delta u] \\[2mm] v_t = -v + u \end{cases} \tag{75}$$

with ε, ε', q, f, d_1, d_3 and ϕ positive parameters and $\overline{u}$ root of the cubic equation

$$u^3 + (q+f-1)u^2 + (\phi - q - fq)u - \phi q = 0. \tag{76}$$

184

In fact, setting

$$\begin{cases} b_{11} = (\varepsilon')^{-1}[-(q+\overline{u}) - d_3\overline{\alpha}_3], \quad b_{12} = (\varepsilon')^{-1}f \\[2mm] b_{13} = -(\varepsilon')^{-1}\overline{w}, \quad b_{21} = \varepsilon^{-1}(q-\overline{u}), \quad b_{22} = 0 \\[2mm] b_{23} = 1 - 2\overline{u} - \overline{w} - d_1\overline{\alpha}_1, \quad b_{31} = 0, \quad b_{32} = -1, \, b_{33} = 1 \end{cases} \tag{77}$$

(75) becomes

$$\begin{cases} w_t = b_{11}w + b_{12}v + b_{13}u \\[2mm] u_t = b_{21}w + b_{23}u \\[2mm] v_t = b_{32}v + b_{33}u \end{cases} \tag{78}$$

and, at least for $q > \overline{u}$, belongs to the case ii).

Acknowledgments

This work has been performed under the auspices of the G.N.F.M. of I.N.D.A. M. and M.I.U.R. (P.R.I.N. 2005): "Nonlinear Propagation and Stability in Thermodynamical Processes of Continuous Media".

References

1. A. Okubo, S.A. Levin, *Diffusion and ecological problems: modern perspectives.* Second Edition. Interdisciplinary Applied Mathematics, vol. 14. Springer-Verlag, New York 2001.
2. J.D. Murray, *Mathematical Biology. I, II* . Third Edition. Interdisciplinary Appl.Math., vol.17. Springer-Verlag, New York 2002.
3. B. Straughan, *The energy method, stability, and nonlinear convection.* Appl. Math. Sci. Ser. vol. 91, Springer-Verlag. Second edition 2004.
4. R.S. Cantrell, C. Cosner, *Spatial Ecology via Reaction-Diffusion Equations.* Wiley 2003.
5. J.N. Flavin, S. Rionero, *Qualitative estimates for partial differential equations: an introduction.* CRC Press, Boca Raton, Florida 1996.
6. S. Senn, *On a nonlinear elliptic eigenvalue problem with Neumann boundary conditions, with an application to population genetics,* Comm. in P.D.E. **8**(11), pp. 1199-1228 (1993).
7. F.R. Gantmacher, *The theory of matrices,* vol. 1, AMS 200, p. 302.
8. S. Rionero, *A rigorous reduction of the L^2-stability of the solutions to a nonlinear binary reaction-diffusion system of P.D.Es. to the stability of the solutions to a linear binary system of O.D.Es.,* Journal of Mathematical Analysis and Application **319**, pp. 372-392 (2006).
9. S. Rionero, *A nonlinear L^2 stability analysis for two species dynamics with dispersal,* Math. Biosc. Eng., **3**, n. 1, pp. 189-204 (2006).
10. S. Rionero, *Diffusion driven stability and Turing effect under Robin boundary data* (To appear)

11. S. Rionero, L^2-*stability of the solutions to a nonlinear binary reaction-diffusion system of P.D.Es.*. Rend.Mat.Acc.Lincei, s.9, **16** (2005).
12. I. Schebesch, H. Engel, *Interacting spiral waves in the Oregonator model of the light-sensitive Belonsov-Zhabotinskii reaction*. Phys. rev. E **60** (6), pp. 6429-6434.
13. S. Lombardo, G. Mulone, M. Trovato, *A general analitic procedure to obtain optimal Liapunov functions in Reaction-Diffusion systems*. Rend. Circolo Matematico di Palermo, serie II, Suppl. 78, pp. 173-185 (2006).

MIXTURE OF GASES WITH MULTI-TEMPERATURE: MAXWELLIAN ITERATION

T. RUGGERI*

*Department of Mathematics
and Research Center of Applied Mathematics (C.I.R.A.M.)
University of Bologna, Via Saragozza 8, 40123 Bologna, Italy
E-mail: ruggeri@ciram.unibo.it*

S. SIMIĆ

*Department of Mechanics, Faculty of Technical Sciences
University of Novi Sad, Trg Dositeja Obradovica 6, 21000 Novi Sad, Serbia
E-mail: ssimic@uns.ns.ac.yu*

*Dedicato ad Antonio Greco,
uno dei miei più cari amici, con grande affetto

In this paper a hyperbolic model is proposed for mixtures of gases which are neither viscous, nor heat-conducting (Eulerian fluids). It is built upon assumption that each constituent obeys it's own temperature. Restrictions to the structure of the model come out from basic principles of extended thermodynamics, i.e. Galilean invariance of balance laws and entropy inequality. Hierarchy of hyperbolic subsystems is recognized, with a single-temperature model as principal subsystem and classical Euler's equations as equilibrium subsystem. Finally, in order to relate this model to classical thermodynamics, a Maxwellian iteration is performed in the case of binary mixture, giving rise to a relation between the difference of non-equilibrium temperatures of constituents and classical fields.

Keywords: mixture of fluids, extended thermodynamics.

1. Introduction

Accurate description of non-equilibrium phenomena is an intricate problem which calls for permanent improvement of mathematical models and thorough analysis of constitutive assumptions. An example is the mixture theory of gases. In classical thermodynamics of irreversible phenomena (TIP) n–component mixture had been described by means of $n + 4$ fields of mass densities ρ_α, $\alpha = 1, \ldots, n$, the mixture velocity $\mathbf{v}$ and the mixture temperature T. On the other hand, within the framework of rational thermodynamics Truesdell[1] proposed a model of homogeneous mixtures. The basis of this approach is expressed as a principle that all properties of the mixture must be mathematical consequences of properties of the constituents. In accordance with this approach Müller[2] developed appropriate constitutive theory for the mixtures. This model described mixture using $4n + 1$ field of densities ρ_α, constituent velocities $\mathbf{v}_\alpha$ and mixture temperature T.

In parallel to the single–temperature (ST) model, a multi–temperature (MT) one was developed within the same context[1], albeit independently of the former. The MT model was described with $5n$ fields of densities, velocities and temperatures of the constituents $(\rho_\alpha, \mathbf{v}_\alpha, T_\alpha)$. The differences between these two models were apparent, in particular with respect to entropy principle and wave propagation, see Refs. 2,3 and references cited therein. Moreover, two theories are quite different also in mathematical sense — their solutions cannot be related in a simple way since MT model, in principle, does not admit a solution of the form $T_1 = \ldots = T_n = T$, even when the initial temperatures of the constituents are equal. Although MT approach could be criticized for physical reasons, it is naturally imbedded into Maxwell's kinetic theory of mixtures[4,5]. This study has an intention to relate these two models using the basic principles of extended thermodynamics[3] and hierarchical structure of hyperbolic subsystems, as explained by Boillat and Ruggeri[6]. These results will facilitate the study of global existence and asymptotic behavior of the solutions. Finally, through the model of binary mixture a relation between the present model and the classical one will be established through the application of Maxwellian iteration procedure. Extensive explanation of results presented in Sects. 2 and 3 can be found in Ref. 7.

2. The multi–temperature model

Following the postulate of rational thermodynamics[1] that each constituent obeys the same balance laws as a single fluid, $5n$ balance laws of mass, momentum and energy for constituents could be used as governing equations for the MT mixture, with appropriate production terms due to mutual interaction of the constituents. However, it is convenient and conceptually adequate to write down also conservation laws of mass, momentum and energy of the whole mixture. As a consequence mathematical model for the MT mixture could be consisted of 5 conservation laws for the mixture as a whole and $5(n-1)$ balance laws for $n-1$ constituents, dropping the balance laws for say n^{th} constituent

$$\frac{\partial \rho}{\partial t} + \mathrm{div}(\rho \mathbf{v}) = 0;$$

$$\frac{\partial (\rho \mathbf{v})}{\partial t} + \mathrm{div}(\rho \mathbf{v} \otimes \mathbf{v} - \mathbf{t}) = \mathbf{0};$$

$$\frac{\partial \left(\frac{1}{2} \rho v^2 + \rho \varepsilon \right)}{\partial t} + \mathrm{div} \left\{ \left(\frac{1}{2} \rho v^2 + \rho \varepsilon \right) \mathbf{v} - \mathbf{t}\mathbf{v} + \mathbf{q} \right\} = 0;$$

$$\frac{\partial \rho_b}{\partial t} + \mathrm{div}(\rho_b \mathbf{v}_b) = \tau_b \tag{1}$$

$$\frac{\partial (\rho_b \mathbf{v}_b)}{\partial t} + \mathrm{div}(\rho_b \mathbf{v}_b \otimes \mathbf{v}_b - \mathbf{t}_b) = \mathbf{m}_b, \qquad (b = 1, 2, \ldots, n-1)$$

$$\frac{\partial \left(\frac{1}{2} \rho_b v_b^2 + \rho_b \varepsilon_b \right)}{\partial t} + \mathrm{div} \left\{ \left(\frac{1}{2} \rho_b v_b^2 + \rho_b \varepsilon_b \right) \mathbf{v}_b - \mathbf{t}_b \mathbf{v}_b + \mathbf{q}_b \right\} = e_b,$$

Here usual notation of continuum thermomechanics has been exploited. Density ρ, (barycentric) velocity $\mathbf{v}$ and internal energy density ε of the mixture, as well as stress tensor $\mathbf{t}$ and flux of internal energy $\mathbf{q}$ are defined as

$$\rho = \sum_{\alpha=1}^{n} \rho_\alpha; \quad \mathbf{v} = \frac{1}{\rho} \sum_{\alpha=1}^{n} \rho_\alpha \mathbf{v}_\alpha; \quad \varepsilon = \frac{1}{\rho} \sum_{\alpha=1}^{n} \rho_\alpha \left(\varepsilon_\alpha + \frac{1}{2} u_\alpha^2 \right);$$

$$\mathbf{t} = \sum_{\alpha=1}^{n} (\mathbf{t}_\alpha - \rho_\alpha \mathbf{u}_\alpha \otimes \mathbf{u}_\alpha); \tag{2}$$

$$\mathbf{q} = \sum_{\alpha=1}^{n} \left\{ \mathbf{q}_\alpha + \rho_\alpha \left(\varepsilon_\alpha + \frac{1}{2} u_\alpha^2 \right) \mathbf{u}_\alpha - \mathbf{t}_\alpha \mathbf{u}_\alpha \right\}.$$

In (2) $\mathbf{u}_\alpha = \mathbf{v}_\alpha - \mathbf{v}$ denotes diffusion velocity with obvious property $\sum_{\alpha=1}^{n} \rho_\alpha \mathbf{u}_\alpha = \mathbf{0}$ and production terms satisfy following relations

$$\sum_{\alpha=1}^{n} \tau_\alpha = 0; \quad \sum_{\alpha=1}^{n} \mathbf{m}_\alpha = \mathbf{0}; \quad \sum_{\alpha=1}^{n} e_\alpha = 0. \tag{3}$$

Equations (1) are the governing equations for thermodynamic processes in MT mixtures. The goal of extended thermodynamics is to determine the $5n$ fields describing the state of the mixture as a whole and the states of $n-1$ constituents. The state of nth constituent can be obtained via $(2)_{1-3}$ and constitutive equations.

2.1. Euler fluids

So far nothing has been told about constitutive assumptions and the structure of production terms. Firstly, our attention will be restricted to so-called Euler fluids — fluids which are neither viscous, nor heat–conducting. This assumption leads to the following equations for stress tensors and heat fluxes

$$\mathbf{t}_\alpha = -p_\alpha \mathbf{I}; \quad \mathbf{q}_\alpha = \mathbf{0}; \quad (\alpha = 1, \ldots, n), \tag{4}$$

where p_α denote partial pressures.

Furthermore, it will be assumed that constituents obey classical thermal and caloric equations of state

$$p_\alpha = \frac{k}{m_\alpha} \rho_\alpha T_\alpha; \quad \varepsilon_\alpha = \frac{p_\alpha}{\rho_\alpha (\gamma_\alpha - 1)}; \quad (\alpha = 1, \ldots, n), \tag{5}$$

k being the Boltzmann constant and m_α is the atomic mass of the α-constituent.

2.2. Galilean invariance

The MT model for the mixture of Euler fluids are a particular example of the system of balance laws

$$\partial_t \mathbf{F}^0 + \partial_i (\mathbf{F}^0 v^i + \mathbf{G}^i) = \mathbf{F}, \tag{6}$$

where $\mathbf{F}^0$, $\mathbf{G}^i$ and $\mathbf{F}$ are densities, non-convective fluxes and production terms, respectively. Principle of relativity requires that equations (6) are invariant with respect to Galilean transformations. As a consequence, see Ruggeri[8], there exists a linear operator $\mathbf{X}(\mathbf{v})$ such that:

$$
\begin{aligned}
\mathbf{F}^0(\mathbf{w}, \mathbf{v}) &= \mathbf{X}(\mathbf{v})\hat{\mathbf{F}}^0(\mathbf{w}), \\
\mathbf{G}^i(\mathbf{w}, \mathbf{v}) &= \mathbf{X}(\mathbf{v})\hat{\mathbf{G}}^i(\mathbf{w}), \\
\mathbf{F}(\mathbf{w}, \mathbf{v}) &= \mathbf{X}(\mathbf{v})\hat{\mathbf{F}}(\mathbf{w}),
\end{aligned}
\tag{7}
$$

where $\mathbf{v}$ is the mixture velocity and $\mathbf{w}$ is a vector of the other field variables and the hat indicates, also in the follows, the corresponding quantity evaluated at zero velocity. These conditions reveal the velocity dependence of production terms in the following way[7]

$$
\begin{aligned}
\tau_b &= \hat{\tau}_b; \\
\mathbf{m}_b &= \hat{\tau}_b \mathbf{v} + \hat{\mathbf{m}}_b; \quad (b = 1, \ldots, n-1) \\
e_b &= \hat{\tau}_b \frac{v^2}{2} + \hat{\mathbf{m}}_b \cdot \mathbf{v} + \hat{e}_b,
\end{aligned}
\tag{8}
$$

while production terms for n^{th} constituent can be derived from Eq. (3).

2.3. *Entropy principle and its restrictions*

Apart from Galilean invariance entropy inequality, i.e. supplementary balance law

$$
\partial_t h^0 + \partial_i(h^0 v^i + \varphi^i) = \Sigma \leq 0,
\tag{9}
$$

imposes another important restriction to the structure of constitutive functions, in particular production terms. Except for the sign h^0, φ^i and Σ are respectively the entropy density, the entropy flux and the entropy production. Since balance laws and entropy inequality are both quasi-linear equations, their compatibility can be achieved through the use of the main field[9,10] $\mathbf{u}'$. Denoting the main field components by

$$
\mathbf{u}' = \frac{\partial h^0}{\partial \mathbf{u}} = \left(\Lambda^\rho, \mathbf{\Lambda}^\mathbf{v}, \Lambda^\varepsilon, \Lambda^{\rho_b}, \mathbf{\Lambda}^{\mathbf{v}_b}, \Lambda^{\varepsilon_b}\right); \quad (b = 1, \ldots, n-1),
\tag{10}
$$

and exploiting the so-called residual inequality

$$
\Sigma = \mathbf{u}' \cdot \mathbf{F} = \hat{\mathbf{u}}' \cdot \hat{\mathbf{F}} = \sum_{b=1}^{n-1} \left(\hat{\Lambda}^{\rho_b}\hat{\tau}_b + \hat{\mathbf{\Lambda}}^{\mathbf{v}_b} \cdot \hat{\mathbf{m}}_b + \hat{\Lambda}^{\varepsilon_b}\hat{e}_b\right) \leq 0,
\tag{11}
$$

one may obtain more explicit structure of internal parts of production terms. Choosing them in a way that (11) becomes a quadratic form, the following expressions are obtained[7]

$$
\hat{\tau}_b = -\sum_{c=1}^{n-1} \varphi_{bc}(\mathbf{w}) \left(\frac{\mu_c - \frac{1}{2}u_c^2}{T_c} - \frac{\mu_n - \frac{1}{2}u_n^2}{T_n}\right);
\tag{12}
$$

$$
\hat{\mathbf{m}}_b = -\sum_{c=1}^{n-1} \psi_{bc}(\mathbf{w}) \left(\frac{\mathbf{u}_c}{T_c} - \frac{\mathbf{u}_n}{T_n}\right); \quad \hat{e}_b = -\sum_{c=1}^{n-1} \theta_{bc}(\mathbf{w}) \left(-\frac{1}{T_c} + \frac{1}{T_n}\right),
$$

190

where $\mu_\alpha = \varepsilon_\alpha - T_\alpha S_\alpha + p_\alpha/\rho_\alpha$, $\alpha = 1,\dots,n$ are chemical potentials of the constituents and $\varphi_{bc}(\mathbf{w})$, $\psi_{bc}(\mathbf{w})$ and $\theta_{bc}(\mathbf{w})$, $b,c = 1,\dots,n-1$ are positive definite matrix functions of the objective quantities that we assume symmetric according with Onsager idea.

3. Principal subsystems and asymptotic behavior

Main field (10) can be used to transform the system of balance laws (6) into symmetric hyperbolic form

$$\partial_t \left(\frac{\partial h'^0}{\partial \mathbf{u}'} \right) + \partial_i \left(\frac{\partial h'^i}{\partial \mathbf{u}'} \right) = \mathbf{F}. \tag{13}$$

We split into two blocks of R^M and R^{N-M} vectors the main field and the production R^N-vectors: $\mathbf{u}' \equiv (\mathbf{v}', \mathbf{w}')$, $\mathbf{F} \equiv (\mathbf{f}(\mathbf{u}'), \mathbf{g}(\mathbf{u}'))$.

Assigned a constant value $\mathbf{w}'_*$ to the second group of variables, the system of the first M equations

$$\partial_t \left(\frac{\partial h'^0(\mathbf{v}', \mathbf{w}'_*)}{\partial \mathbf{v}'} \right) + \partial_i \left(\frac{\partial h'^i(\mathbf{v}', \mathbf{w}'_*)}{\partial \mathbf{v}'} \right) = \mathbf{f}(\mathbf{v}', \mathbf{w}'_*) \tag{14}$$

is called[6] a *principal subsystem* of (13).

It is also symmetric hyperbolic and obey the sub-characteristic condition and sub-entropy law[6]. A special class of systems (13) is the one for which $\mathbf{f}(\mathbf{u}') \equiv \mathbf{0}$. In this case, for $\mathbf{w}'_* = \mathbf{0}$ system (14) appears to be an equilibrium subsystem in which entropy production $-\Sigma|_E$ vanishes and attains its minimum.

3.1. *Principal subsystems in MT mixture model*

In the case of MT mixture an appropriate hierarchy of subsystems could be obtained using procedure described above. Most interesting cases are listed below[7]:

Case 1. A single-temperature model is a principal subsystem of the MT model. When $\Lambda^{\varepsilon b} = 0$, $b = 1,\dots,n-1$ is put, one obtains

$$T_1 = \dots = T_n = T, \tag{15}$$

and balance laws for the energy should be dropped from the model. At the same time all the constitutive functions have to be evaluated in accordance with this restriction.

Case 2. If along with $\Lambda^{\varepsilon b} = 0$, $\mathbf{\Lambda}^{\mathbf{v} b} = 0$, $b = 1,\dots,n-1$ is valid, the following restriction is obtained

$$\mathbf{u}_1 = \dots = \mathbf{u}_n = \mathbf{0}, \tag{16}$$

which leads to a single temperature model where balance laws for momenta have to be dropped. This kind of model is widely accepted for modeling of detonation phenomena.

Case 3. If yet another condition is imposed $\Lambda^{\rho b}$, $b = 1, \ldots, n-1$, the following restriction is obtained

$$\mu_1 = \ldots = \mu_n = \mu, \tag{17}$$

giving rise to equilibrium densities ρ_α of the constituents. Since balance laws for the masses have to be dropped, one arrives to the classical Euler's system of equations of gas dynamics, here representing an equilibrium subsystem.

3.2. *Qualitative analysis*

The MT model for mixtures is a particular case of a system of balance laws (6) which is dissipative due to the presence of the productions that satisfy the entropy principle. On the other hand the model is of mixed type since five equations are conservation laws while the rest of the system is consisted of true balance laws

$$\mathbf{F}(\mathbf{u}) = \begin{pmatrix} 0 \\ \mathbf{g}(\mathbf{u}) \end{pmatrix}.$$

For this class of systems the coupling condition of Shizuta and Kawashima[11] (K-condition) can ensure the global existence of smooth solutions, meaning that dissipation in balance laws prevails the hyperbolicity of conservation laws. Actually, the K-condition reads: *In the equilibrium manifold any characteristic eigenvector is not in the null space of* $\nabla \mathbf{F}$, i.e.

$$\nabla \mathbf{F} \cdot \mathbf{d}^{(i)} \Big|_E \neq 0 \quad \forall i = 1, \ldots, N. \tag{18}$$

where $\mathbf{d}^{(i)}$ are the right-eigenvectors of the hyperbolic system (6):

$$(\mathbf{A}_n - \lambda \mathbf{I}) \, \mathbf{d} = 0, \tag{19}$$

where $\mathbf{A}_n = \mathbf{A}^i n_i$, $\mathbf{A}^i = \nabla \mathbf{F}^i$, $\mathbf{u} = \mathbf{F}^0$, $\nabla = \partial/\partial \mathbf{u}$ and E stands for the equilibrium state.

It has been proven recently that if the system of balance laws (6) is endowed with a convex entropy law, and it is dissipative, then the K-condition becomes a sufficient condition for the existence of global smooth solutions provided that the initial data are sufficiently smooth.[12,13] Furthermore, Ruggeri and Serre[14] proved the stability of constant states in one-dimensional case.

After an extensive calculation[7] it has been shown that K–condition is satisfied for the MT model of mixtures and that, according to above mentioned results, the following statement is valid:

If the initial data of the MT model are perturbations of equilibrium state, smooth solutions exist for all time and tends to the equilibrium constant state.

4. Binary mixtures and Maxwell Iteration

The multi–temperature model for mixtures, proposed within the context of extended thermodynamics, possesses desirable properties such as hyperbolicity, symmetric form of governing equations ensuring well-posedness of initial-value

192

problems and global existence of smooth solutions. It is therefore quite natural to investigate its relation to the mixture model proposed in classical thermodynamics. There is a substantial difference between classical and extended approach. Diffusion flux and flux of internal energy are given by constitutive equations in classical thermodynamics, whereas in extended one they represent variables determined by the set of differential equations. Moreover, new model have the temperatures of constituents as additional independent variables whose evolution is governed by the balance laws for energy. To get a first impression about relation between classical and extended MT model a special case of binary mixture will be investigated.

In the case of binary non–reacting mixture of Euler fluids mathematical model could be written in the form[7]

$$\frac{\partial \rho}{\partial t} + \mathrm{div}\,(\rho \mathbf{v}) = 0;$$

$$\frac{\partial \rho \mathbf{v}}{\partial t} + \mathrm{div}\left(\rho \mathbf{v} \otimes \mathbf{v} + p\mathbf{I} + \frac{1}{\rho c(1-c)}\mathbf{J} \otimes \mathbf{J}\right) = \mathbf{0};$$

$$\frac{\partial \left(\frac{1}{2}\rho v^2 + \rho \varepsilon\right)}{\partial t} + \mathrm{div}\left\{\left(\frac{1}{2}\rho v^2 + \rho \varepsilon + p\right)\mathbf{v} + \left(\frac{\mathbf{v}\cdot\mathbf{J}}{\rho c(1-c)} + \frac{1}{\alpha}\right)\mathbf{J}\right\} = 0;$$

$$\frac{\partial(\rho c)}{\partial t} + \mathrm{div}\,(\rho c \mathbf{v} + \mathbf{J}) = 0; \qquad\qquad (20)$$

$$\frac{\partial(\rho c \mathbf{v} + \mathbf{J})}{\partial t} + \mathrm{div}\left\{\rho c \mathbf{v} \otimes \mathbf{v} + \frac{1}{\rho c}\mathbf{J} \otimes \mathbf{J} + \mathbf{v} \otimes \mathbf{J} + \mathbf{J} \otimes \mathbf{v} + \nu\mathbf{I}\right\} = \mathbf{m}_1;$$

$$\frac{\partial \left(\frac{1}{2}\rho c \left(\mathbf{v} + \frac{\mathbf{J}}{\rho c}\right)^2 + \rho c e\right)}{\partial t} +$$

$$\mathrm{div}\left\{\left(\frac{1}{2}\rho c \left(\mathbf{v} + \frac{\mathbf{J}}{\rho c}\right)^2 + \rho c e + \nu\right)\left(\mathbf{v} + \frac{\mathbf{J}}{\rho c}\right)\right\} = e_1,$$

where $c = \rho_1/\rho$, $\nu = p_1$ and $e = \varepsilon_1$. Diffusion flux vector is defined as

$$\mathbf{J} = \alpha \mathbf{q} = \rho_1 \mathbf{u}_1 = -\rho_2 \mathbf{u}_2,$$

where $\alpha = 1/(g_1 - g_2)$ is thermal inertia of the mixture with $g_\beta = \varepsilon_\beta + (p_\beta/\rho_\beta) + u_\beta^2/2$, $\beta = 1, 2$, being non-equilibrium enthalpies.

Thermodynamic process in binary mixture, governed by equations (20), is described by the field variables $(\rho, \mathbf{v}, T, c, \mathbf{J}, \Theta)$, with $\Theta = T_2 - T_1$. Therefore, $\mathbf{J}$ and Θ can be viewed as non-equilibrium variables in the sense of extended thermodynamics. Source terms $\mathbf{m}_1$ and e_1 are determined in accordance with general results (12)

$$\mathbf{m}_1 = \hat{\mathbf{m}}_1 = -\psi_{11}\left(\frac{1}{cT_1} - \frac{1}{(1-c)T_2}\right)\frac{\mathbf{J}}{\rho};$$

$$\qquad\qquad (21)$$

$$\hat{e}_1 = \theta_{11}\frac{\Theta}{T_1 T_2}; \quad e_1 = \hat{\mathbf{m}}_1 \cdot \mathbf{v} + \hat{e}_1.$$

To reveal the relation between extended and classical model, a formal iterative scheme known as Maxwellian iteration will be applied.[3] The first iterates $\mathbf{J}^{(1)}$ and $\Theta^{(1)}$ are calculated from the right–hand sides of Eqs. $(20)_{5,6}$ by putting "zeroth" iterates — equilibrium values $\mathbf{J}^{(0)} = \mathbf{0}$ and $\Theta^{(0)} = 0$ on the left–hand sides. In the next step second iterates $\mathbf{J}^{(2)}$ and $\Theta^{(2)}$ are obtained from the right–hand sides of the same Eqs. by putting first $\mathbf{J}^{(1)}$ and $\Theta^{(1)}$ on their left–hand sides, an so on. Here, only the first iterates will be derived.

By applying the Maxwellian iteration scheme to Eq. $(20)_5$, after some simple manipulations with the use of Eqs. $(20)_{1,4}$ one obtains the first iterate of diffusion flux

$$\mathbf{J}^{(1)} = \tau_J \left(c\, \mathrm{grad} p - \mathrm{grad}\nu \right); \quad \tau_J = \frac{\rho c(1-c)}{\psi_{11}}, \tag{22}$$

where τ_J stands for the relaxation time of diffusion. In expanded form this relation reads

$$\mathbf{J}^{(1)} = k\,\tau_J \left\{ c(1-c) \left(\frac{1}{m_1} - \frac{1}{m_2} \right) (T\,\mathrm{grad}\rho + \rho\,\mathrm{grad}T) + \frac{\rho}{m(c)} T \mathrm{grad} c \right\}.$$

This equation can be recast into the form

$$\mathbf{J}^{(1)} = L_{11}\,\mathrm{grad}\left(\frac{\mu_2 - \mu_1}{T} \right) + L\,\mathrm{grad}\left(\frac{1}{T} \right),$$

well-known from TIP as Fick Law[3,15] . Phenomenological coefficients L_{11} and L are related to our field variables in the following way

$$L_{11} = \frac{\rho^2 c^2 (1-c)^2 T^2}{\psi_{11}}$$

$$L = -k\,\frac{\rho^2 c^2 (1-c)^2 T^3}{\psi_{11}} \left\{ \frac{\gamma_2}{\gamma_2 - 1}\frac{1}{m_2} - \frac{\gamma_1}{\gamma_1 - 1}\frac{1}{m_1} \right\}.$$

Treating Eq. $(20)_6$ in the same manner one obtains

$$\Theta^{(1)} = \tau_T \left(\frac{dT}{dt} + (\gamma_1 - 1)T\mathrm{div}\mathbf{v} \right); \quad \tau_T = \frac{k}{m_1(\gamma_1 - 1)}\frac{\rho c T^2}{\theta_{11}},$$

where τ_T stands for relaxation time of non-equilibrium temperatures. Taking into account Eqs. $(20)_{3,4}$ and constitutive functions last equation can be reduced to

$$\Theta^{(1)} = L_\theta \left(\gamma_1 - \gamma_2 \right) \mathrm{div}\mathbf{v}. \tag{23}$$

where the corresponding phenomenological coefficient L_θ reads

$$L_\theta = \frac{k\rho c(1-c)T^3}{\theta_{11}(m_2(\gamma_2 - 1)c + m_1(\gamma_1 - 1)(1-c))}. \tag{24}$$

The equation (23), which seems to appear for the first time, expresses the temperature difference, here in the role of thermodynamic flux, as a linear function of divergence of velocity, i.e. thermodynamic force, as it is usual in non-equilibrium thermodynamics.

194

From (23) *it is very interesting to observe that in the classical TIP approach no possible difference of temperature between components arise if the ratio of the specific heats are the same.*

Equation (23) is a simplified version of general expression which can be obtained in the case of multi-component reacting mixtures. Being scalar quantities, the differences of temperatures $\Theta_b = T_n - T_b$ could also depend, in the first iteration, on so–called chemical "forces" which reflect the structure of chemical reactions taking place during the process.

Comparing our equations (1) with the ones obtained by the kinetic theory, we are able to evaluate the coefficients θ_{11} and ψ_{11}:

$$\psi_{11} = 2T\frac{m_1 m_2}{m_1 + m_2}\Gamma'_{12}; \qquad \theta_{11} = 3kT^2\frac{m_1 m_2}{(m_1 + m_2)^2}\Gamma'_{12}$$

where Γ'_{12} represents the number of collision per density volume[16] .

Acknowledgment: This paper was supported in part (T.R.) by fondi MIUR Progetto di interesse Nazionale *Problemi Matematici Non Lineari di Propagazione e Stabilità nei Modelli del Continuo* Coordinatore T. Ruggeri, by the GNFM-INdAM, and (S.S.) by the Ministry of Science and Environmental Protection of Serbia within the project *Contemporary Problems of Mechanics of Deformable Bodies.*

References

1. C. Truesdell, *Rational Thermodynamics* (McGraw-Hill Series in Modern Applied Mathematics, McGraw-Hill, New York, 1969).
2. I. Müller, *Thermodynamics* (Pitman Publ. Co., London, 1985).
3. I. Müller, T. Ruggeri, *Rational Extended Thermodynamics*, 2nd ed. (Springer Tracts in Natural Philosophy **37**, Springer-Verlag, New York, 1998).
4. S.C. Chapman, T.G. Cowling, *The Mathematical Theory of Non–Uniform Gases* (Cambridge University Press, London, 1961).
5. J.M. Burgers, *Flow Equations for Composite Gases* (Academic Press, New York, 1969).
6. G. Boillat, T. Ruggeri, *Arch. Rational Mech. Anal.* **137**, 305 (1997).
7. T. Ruggeri, S. Simić, *Math. Meth. Appl. Sci.* (in press). Pubblished online DOI: 10.1002/mma.813 (2006).
8. T. Ruggeri, *Continuum Mech. Thermodyn.* **1**, 3 (1989).
9. G. Boillat, *C.R. Acad. Sc. Paris* **278** A, 909 (1974).
10. T. Ruggeri, A. Strumia, *Ann. Inst. H. Poincaré* **34** A, 65 (1981).
11. Y. Shizuta, S. Kawashima, *Hokkaido Math. J.* **14**, 249 (1985).
12. B. Hanouzet, R. Natalini, *Arch. Rational Mech. Anal.* **163**, 89 (2003).
13. W.A. Yong, *Arch. Rational Mech. Anal.* **172**, 247 (2004).
14. T. Ruggeri, D. Serre, *Quart. Appl. Math.* **62**(1), 163 (2004).
15. S.R. de Groot, P. Mazur, *Non-equilibrium thermodynamics* (North–Holland Pub. Co., Amsterdam, 1962).
16. T.K. Bose, *High Temperature Gas Dynamics* (Springer, Berlin, 2003).

BIFURCATION ANALYSIS OF SEQUENCE OF MAGNETIC ISLAND EQUILIBRIA AND SPONTANEOUS GENERATION OF ZONAL FLOWS

C. Tebaldi

Department of Mathematics and Burning Plasma Research Group,
Politecnico of Torino
Corso Duca degli Abruzzi 24, 10129 Torino, Italy
E-mail: claudio.tebaldi@polito.it

L. Margheriti

Department of Mathematics, University of Messina
Salita Sperone 31, 98166 Messina, Italy
E-mail: margheriti@mat520.unime.it

D. Grasso

Department of Energetics and Burning Plasma Research Group,
Politecnico di Torino
Corso Duca degli Abruzzi 24, 10129 Torino, Italy
E-mail: daniela.grasso@infm.polito.it

The sequence of equilibria of two dimensional reduced magnetohydrodynamics has been studied as a function of the aspect ratio ϵ as a control parameter of the magnetic shear for different equilibrium magnetic fields. A threshold in the magnetic shear parameter is found, below which the small island reconnected equilibrium disappears by tangent bifurcation, while above it undergoes a further symmetry-breaking bifurcation leading to equilibria with enhanced kinetic energy in the form of zonal flow. The spontaneous development of zonal flows may ameliorate transport and improve confinement in tokamak devices.

1. Introduction

In magnetic fusion experiments tearing modes[1] are often responsible for degraded plasma confinement and are seen as a potential threat to the successful operation of burning plasma experiments. These modes are the result of a spontaneous magnetic reconnection process, commonly observed in magnetically confined plasmas, both in space and in laboratory experiments.[2,3] The result is a change of the topology of the magnetic field, with appearance of magnetic islands, and the conversion of magnetic energy into plasma kinetic energy and heat in a time much shorter than the global resistive diffusion time.

In ref.[4] it has been shown that magnetic islands can develop zonal flows. In a low-β (= kinetic pressure/magnetic pressure) plasma, a tiny percentage of the magnetic energy transformed into kinetic energy is sufficient to lead to a significant generation of velocity fields. Although the development of magnetic islands is generally expected to have a detrimental effect on confinement, the spontaneous development of zonal flows may ameliorate transport. Indeed, sheared velocity layers are capable of suppressing turbulence induced by electrostatic drift instabilities, widely believed to be responsible for anomalous transport in tokamak plasmas.[5]

In this work, we show that the generation of zonal flows is related to the combined effect of a subcritical symmetry breaking bifurcation and a tangent bifurcation in the sequence of the equilibria with magnetic island. In order to obtain the detailed bifurcation diagram the resolution of the fixed-point code, essential to track the equilibria, had to be increased respect to ref.[4]

This work is organized as follows. Sec. 2 is devoted to the illustration of the MHD model and of the numerical techniques. In Sec. 3 we consider the class of symmetric equilibria $\psi_e^\alpha(x) = \psi_0^\alpha/(\cosh^2 \alpha x)$ with no motion and we analyse the complete bifurcation diagram in the case $\alpha = 1$ and $\alpha = 2$. Conclusions and a discussion of possible developments are given in Sec. 4.

2. The MHD model and numerical techniques

We consider a two-dimensional incompressible plasma obeying the reduced resistive magneto-hydrodynamics (RRMHD) equations[10]

$$\partial_t U + [\phi, U] = [J, \psi] + \mu \nabla^2 U \tag{1}$$

$$\partial_t \psi + [\phi, \psi] = -\eta(J - J_e) \tag{2}$$

This two-dimensional model is a great simplification, which is appropriate when the unreconnected configuration has at least one ignorable coordinate and a strong guide field along the ignorable direction is present. The equations are defined on a two-dimensional domain with coordinates x and y. With reference to the magnetic geometry of a tokamak, x can be thought of as a radial coordinate and y as a poloidal coordinate. The third direction is considered ignorable. The model equations describe the evolution of a plasma vorticity $U = \nabla^2 \phi$, where ϕ is a stream function, and of the magnetic flux function ψ associated with the magnetic field in the plasma (a constant magnetic field is assumed in the ignorable direction). The other fields are the current density $J = -\nabla^2 \psi$ and a driving current density J_e, associated with equilibrium. Moreover for any two fields A and B, $[A, B] \equiv \partial_x A \partial_y B - \partial_y A \partial_x B$, so that $[\phi, \cdot] = \underline{v} \cdot \nabla$ is the usual advection operator.

Lengths are normalized to a macroscopic length L, which is either a measure of the size of the system or of the scale length of the equilibrium magnetic field. Times are normalized to the Alfvèn time $\tau_A = L/v_A$, where $v_A = B_\rho/\rho^{1/2}$ is the poloidal Alfvèn speed associated with the equilibrium field B_ρ (ρ is the mass density).

The dissipation is measured by the viscosity μ and by the resistivity η, which in these units are respectively the inverse of the Reynolds number R, $\mu = 1/R$, and of the Lundquist number S, $\eta = 1/S$.

The domain is taken to be a square box (slab) $[-L_x, L_x] \times [-L_y, L_y]$, where the normalized lengths are of order one. It is convenient to take $L_x = \pi$, $L_y = \pi/\epsilon$, where the slab aspect ratio ϵ has been introduced. The boundary conditions are taken periodic in both directions.

The model is controlled by three dimensionless parameters, ϵ, S and the magnetic Prandtl number $Pr = S/R$.

In order to solve the system of Eqs. 1-2 a spectral decomposition is adopted for the unknowns, choosing the eigenfunctions of the Laplacian as the complete orthogonal set for the expansion. One has

$$(\phi, \psi) = \sum_{\mathbf{k}} (\phi_{\mathbf{k}}, \psi_{\mathbf{k}}) e^{i\mathbf{k}\cdot\mathbf{x}} \qquad \mathbf{k} = (l, m\epsilon) \quad l, m \quad \text{integers} \tag{3}$$

We truncate the expansion to a finite set L of $2N$ wave vectors ("modes") such that if $\mathbf{k}$ belongs to L also $-\mathbf{k}$ belongs to L. This gives $4N$ ordinary differential equations for $4N$ real unknowns. Moreover, a $2N$ invariant subspace exists, characterized by imaginary amplitudes for the magnetic and velocity fields, which allows to reduce the system to $2N$ real unknowns. The set L is constructed, starting from a "ball" around the origin with N=364 and adding modes in a slab centered at $m = 0$. It has to be noted that because of the spatial localization the computational problem becomes much heavier than in.[22,23] The resolution has been increased also with respect to,[4] in fact N up to $N = 985$ had to be considered, in order to describe the full details of the complicated bifurcation sequence described in the next section.

A suitable tool to find the equilibria is Newton's method, used in connection with the theorems of bifurcation theory.[20] It allows to find also unstable equilibria, which are essential to obtain the bifurcation diagram and then to fully understand the dynamics in the nonlinear regimes. Furthermore, it avoids the difficulty, encountered by the initial value approach, that the time scales can become extremely long, especially near the bifurcations. Finally this method is an efficient way to track the sequence of equilibria when parameters are varied, even if computational effort strongly increases with the dimensionality of the problem.

Since the model equations are supplemented with periodic boundary conditions, a suitable initial value code is a spectral code, which advances in time the Fourier amplitudes of the relevant fields. In some cases a direct truncation of the model equations to the relevant degrees of freedom was used either to compute stable equilibria or to study transients.

3. Unreconnected equilibria with localized magnetic shear

In order to localize magnetic shear around $x = 0$, the system has been considered for the class of equilibria with no motion

$$\psi_e^\alpha(x) = \frac{\psi_0^\alpha}{\cosh^2 \alpha x} \qquad \alpha \in R \tag{4}$$

where ψ_0^α is a normalization constant. ψ_e^α retains the main qualitative characteristics of the flux $\psi_e(x) = \cos x$, giving a zero of the magnetic field in the origin and similar monotonic properties, and its derivatives go to zero at infinity. Even if the domain $[-\pi, \pi]$ is considered, the magnetic shear is very small in a neighbor, increasing with α, of the boundaries. The stability boundary of the reference equilibrium can be obtained by linear theory.[21] We can write

$$\phi(x,y,t) = \tilde{\phi}(x)e^{-i\omega t + iky} \tag{5}$$

$$\psi(x,y,t) = \psi_e^\alpha(x) + \tilde{\psi}(x)e^{-i\omega t + iky} \tag{6}$$

where $k = m\epsilon$, with m integer. The linearized version of Eq. 1 is given by

$$\omega \nabla^2 \tilde{\phi} = -k(\partial_x \psi_e^\alpha \nabla^2 \tilde{\psi} - \tilde{\psi}\partial_x^3 \psi_e^\alpha) + i\mu \nabla^4 \tilde{\phi} \tag{7}$$

where $\nabla^2 \equiv \partial_x^2 - k^2$. The equilibrium becomes instable when ω^2 goes through zero, then the stability boundary is obtained by solving the linearized equation for $\omega^2 = 0$. In the case $\mu = 0$ the linearized equation is

$$\frac{2\alpha \sin \alpha x}{\cosh^3 \alpha x}[\partial_x^2 \tilde{\psi} + (8\alpha^2 - k^2 - 12\alpha^2 \tanh^2 \alpha x)\tilde{\psi}] = 0 \tag{8}$$

This equation can be solved by asymptotic matching. Far from the $x = 0$ line, where the magnetic field vanishes, the solution is approximated by the solution of the outer equation:

$$\partial_x^2 \tilde{\psi}_{out} + (8\alpha^2 - k^2 - 12\alpha^2 \tanh^2 \alpha x)\tilde{\psi}_{out} = 0 \tag{9}$$

As it's usually done, the stability condition is expressed in terms of Δ', the jump in the logarithmic derivative of ψ_{out}. In this case

$$\Delta' = 2\alpha \left(\frac{6k_1^2 - 9}{k_1(k_1^2 - 4)} - k_1 \right) \tag{10}$$

Instability occurs when $\Delta' > 0$, i.e. when $k < \alpha\sqrt{5}$. $k = m\epsilon$, then instability occurs for $m = 1$ when $\epsilon < \epsilon_0 = \alpha\sqrt{5}$.

3.1. *Bifurcation diagram for $\alpha = 1$*

In this paper the equilibrium for $\alpha = 1$ has been considered:

$$\psi_e^1(x) = \frac{\psi_0^1}{\cosh^2 x} \qquad x \in [-\pi, \pi] \tag{11}$$

This equilibrium has been approximated with its expansion in Fourier series truncated to six terms, which is found to be a good representation of $\psi_e^1(x)$. This choice allows the use of periodic boundary conditions and therefore the use of the spectral code described in Sec. 2. Because of the strong localization of the shear in the unreconnected equilibrium, the modal structure, $N = 348$ for the sinusoidal case, was considered up to $N = 721$ in order to check independence of the results obtained from the truncation procedure.

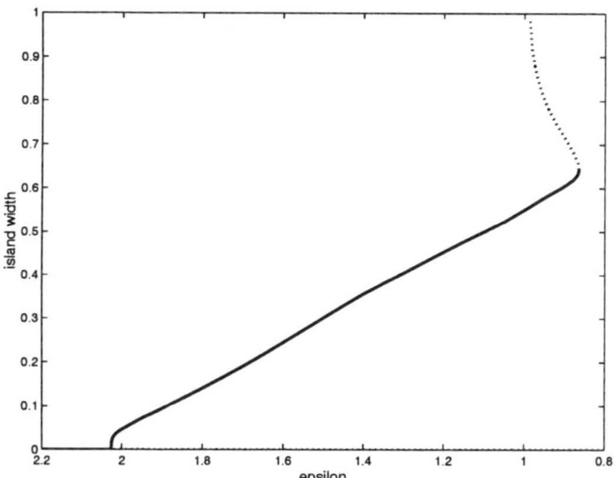

Fig. 1.   Normalized island width w for the equilibria P_0, P and Q^* versus ϵ. Solid lines denote stability, dotted lines instability.

The bifurcation diagram is shown in Fig. 1, where the island width w of the equilibria is plotted against ϵ. A symmetry breaking of the unreconnected equilibrium P_0 (with $w = 0$) at $\epsilon_c = 2.026 < \epsilon_0$ gives rise to a stable equilibrium P with a small island, that disappears at $\epsilon_P = 0.865$ by a tangent bifurcation with an unstable equilibrium Q^*. For values of ϵ less than ϵ_P the system jumps to an equilibrium Q with a macroscopic island. The system shows a second tangent bifurcation for $\epsilon_Q = 0.988$, involving Q and Q^*.

Also with this choice of ψ_e, when the control parameter ϵ is near the threshold of the symmetry-breaking bifurcation ($2.024 \geq \epsilon \geq 2.008$), the island width has a square-root dependence of the amplitude on the departure from ϵ_c, while for $2.005 \geq \epsilon \geq 0.880$, the dependence of w on ϵ is linear.

Fig. 2 shows the contour plots of ψ and ϕ for the three equilibria P, Q^* and Q at a value of ϵ just above ϵ_P. It can be observed that the velocity field is organized in vortices.

The bifurcation diagram is very similar to the one obtained for $\psi_e(x) = \cos x$ in,[22] but the second tangent bifurcation happens for equilibria with island width comparable with the size of the domain. As regards the velocity field, it is more localized around the origin, as it can be expected.

200

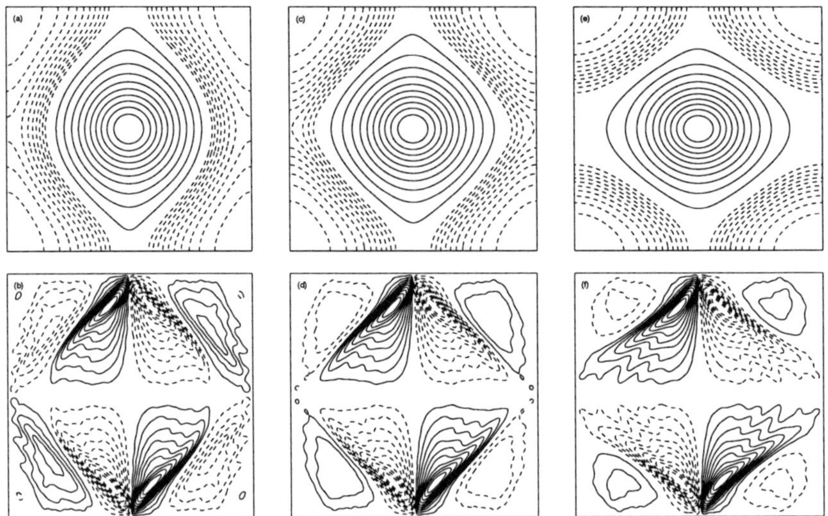

Fig. 2. Contour plots of ψ and ϕ at $Pr = 0.2$, $\alpha = 1$ and $\epsilon = 0.89$ for: P, a) and b), Q^*, c) and d), and Q, e) and f).

These results, obtained for $Pr = 0.2$, seem to be rather insensitive to changes of Prandtl number. In particular, we have considered the cases $Pr = 0.3$ and $Pr = 0.4$ and found the same qualitative behavior of the bifurcation diagram and the same topology of the fields. The values of ϵ_c and ϵ_P for $Pr = 0.3$ and $Pr = 0.4$ are very near to the values found for $Pr = 0.2$. In particular, $\epsilon_c = 2.016$ for $Pr = 0.3$ and $\epsilon_c = 2.008$ for $Pr = 0.4$ (we recall that $\epsilon_c = 2.026$ for $Pr = 0.2$), while $\epsilon_P = 0.865$ for all the three considered values of Prandtl number. Also the island width w_P at the tangent bifurcation is essentially unchanged, $w_P = 0.64$, for $Pr = 0.2$, $Pr = 0.3$, $Pr = 0.4$.

Finally, the behavior of the stability threshold ϵ_c of the symmetric equilibrium has been studied in the limit of zero viscosity and it has been found that $\epsilon_c \longrightarrow \epsilon_0$ for $Pr \longrightarrow 0$, according with the above analytic calculations. Fig.6 represents distance between ϵ_c and ϵ_0 versus Prandtl number. As expected, the role of viscosity is to shift the stability threshold from ϵ_0. A detailed comparison with the scaling reported in[22] , as well as the role of α, has to be investigated.

3.2. Bifurcation diagram for $\alpha = 2$

We have also considered the unreconnected equilibrium for $\alpha = 2$

$$\psi_e^2(x) = \frac{\psi_0^2}{\cosh^2 2x} \qquad x \in [-\pi, \pi] \tag{12}$$

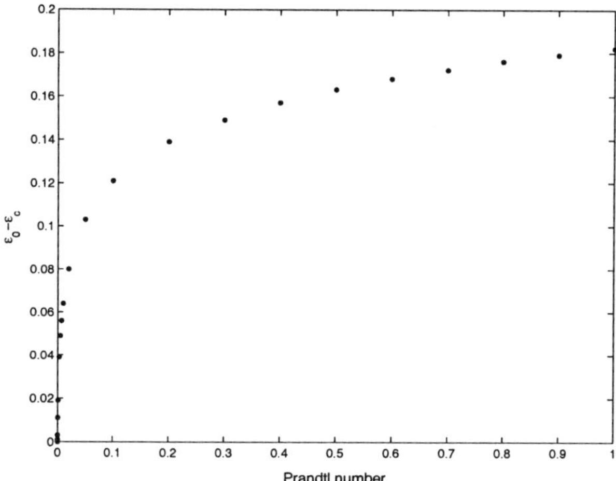

Fig. 3. Distance between ϵ_c and ϵ_0 versus Prandtl number.

This equilibrium has been approximated with an expansion in Fourier series truncated to 21 terms because of the stronger localization of the magnetic shear of the unreconnected equilibrium at $x = 0$ and the modal structure has been taken up to $N = 985$. The results have been found unaffected by the truncation.

The unreconnected equilibrium becomes unstable at $\epsilon_c = 4.162$. Also in this case the destabilization of the unreconnected equilibrium happens through a symmetry breaking bifurcation that originates an equilibrium with a small magnetic island. The equilibrium with small island P doesn't disappear, but becomes unstable at $\epsilon_P = 2.437$ because of a subcritical symmetry breaking bifurcation, involving an unstable equilibrium Q^* with approximately the same island width of P. Because of this fact, the plot w versus ϵ is not an appropriate bifurcation diagram (Fig. 4). Inspection of the kinetic energy for the equilibria shows remarkable differences, therefore kinetic energy is a suitable quantity to represent the bifurcation diagram (Fig. 5).
Q^* disappears at $\epsilon_Q = 2.451$ by a tangent bifurcation with a stable equilibrium Q, then it exists only in a very small range of the parameter ϵ. In this range, which increases when Prandtl number, and then viscosity, decreases (Fig. 6), there are two stable equilibria, P and Q, both with a small island, but with remarkable differences. In fact the pitchfork bifurcation is related to a symmetry breaking in the system: Q and Q^* do not have the same symmetry properties of P. This phenomenon is very relevant in the velocity field, which is not organized anymore for vortices but transport across y-boundaries takes place, along the border of the magnetic island. Furthermore, as can be noted from Fig. 5, the equilibrium Q has higher kinetic energy, which rapidly increases when ϵ decreases from the bifurcation value. Contour plots of ψ and Φ at $\epsilon = 2.44$ for P and at $\epsilon = 2.435$ for Q equilibria are reported in fig. 7. Here, we can appreciate that, while the

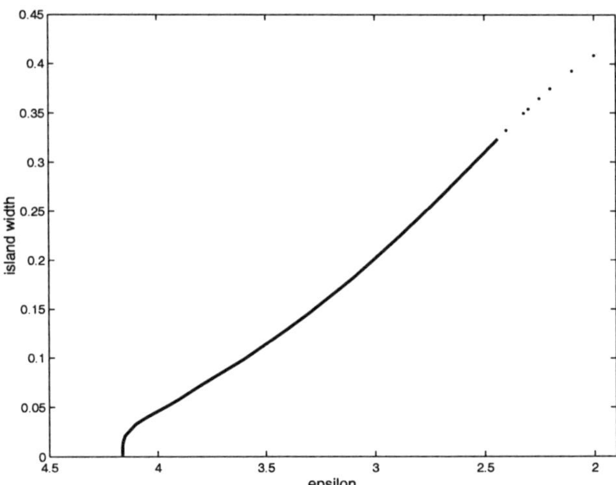

Fig. 4.   Normalized island width w for the equilibrium P versus ϵ. Solid line denotes stability of the symmetric equilibrium.

magnetic islands is essentially the same for the two equilibria P and Q, in the case of P equilibria the velocity field has developed a zonal flow. Finally, we point out that also for $\alpha = 2$, i.e. for higher value of the gradient in the origin, a stable equilibrium with small island disappears by a tangent bifurcation, but this equilibrium is different from the one for $\alpha = 1$, in particular for the kinetic energy.

3.3. *Symmetries and invariant subspaces*

For all the detected equilibria, ψ and ϕ have the symmetry properties

$$\psi(-x, -y, t) = \psi(x, y, t) \tag{13}$$

$$\phi(-x, -y, t) = \phi(x, y, t) \tag{14}$$

which define an invariant subspace characterized by real amplitudes for ψ and ϕ or equivalently imaginary amplitudes for the magnetic and velocity fields (Sec. 2). In this invariant subspace ψ and ϕ are written as

$$\psi(\underline{x}, t) = 2 \sum_{\underline{k} \in L'} \psi_{\underline{k}} \cos \underline{k} \cdot \underline{x} \tag{15}$$

$$\phi(\underline{x}, t) = 2 \sum_{\underline{k} \in L'} \phi_{\underline{k}} \cos \underline{k} \cdot \underline{x} \tag{16}$$

where L' includes the vectors $\underline{k} = (k_x, k_y)$ with $k_x > 0$ and $\underline{k} = (0, k_y)$ with $k_y > 0$. $\psi_{\underline{k}}$ and $\phi_{\underline{k}}$ are the real amplitudes of ψ and ϕ.

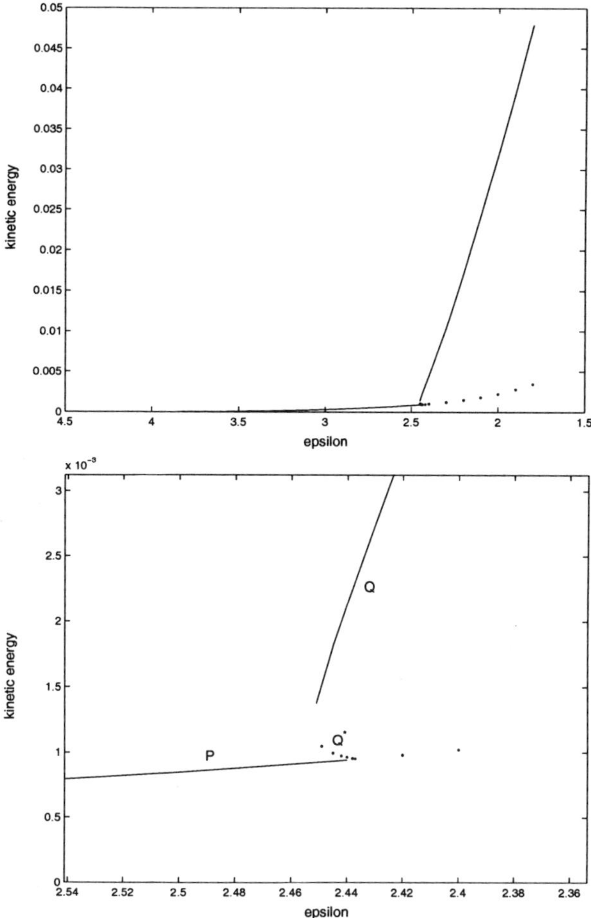

Fig. 5. Kinetic energy for the equilibria P, Q, Q^* versus ϵ. Solid lines denote stability, dotted lines instability.

The equilibrium P has also the symmetry properties

$$\psi(-x, y, t) = \psi(x, -y, t) = \psi(x, y, t) \tag{17}$$

$$\phi(-x, y, t) = \phi(x, -y, t) = -\phi(x, y, t) \tag{18}$$

which define an invariant subspace included in the above one and characterized by the conditions

$$\psi_{(k_x, -k_y)} = \psi_{(k_x, k_y)} \qquad \forall(k_x, k_y), k_x \neq 0, k_y \neq 0 \tag{19}$$

$$\phi_{(k_x, -k_y)} = -\phi_{(k_x, k_y)} \qquad \forall(k_x, k_y), k_x \neq 0, k_y \neq 0 \tag{20}$$

$$\phi_{(0, k_y)} = \phi_{(k_x, 0)} = 0 \qquad \forall k_x, k_y \tag{21}$$

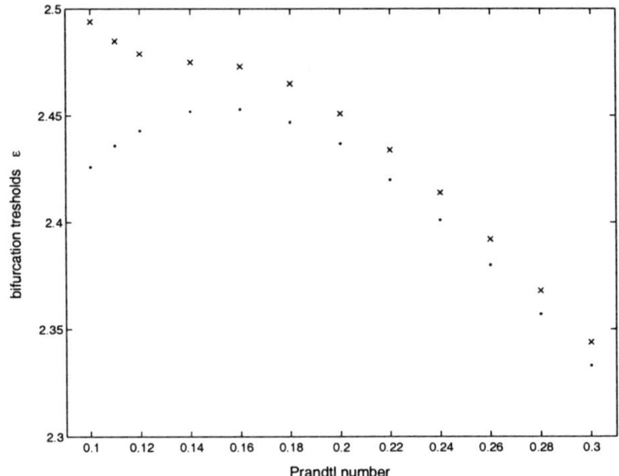

Fig. 6.   Bifurcation thresholds versus Prandtl number. Crosses denote tangent bifurcation, points denote pitchfork bifurcation.

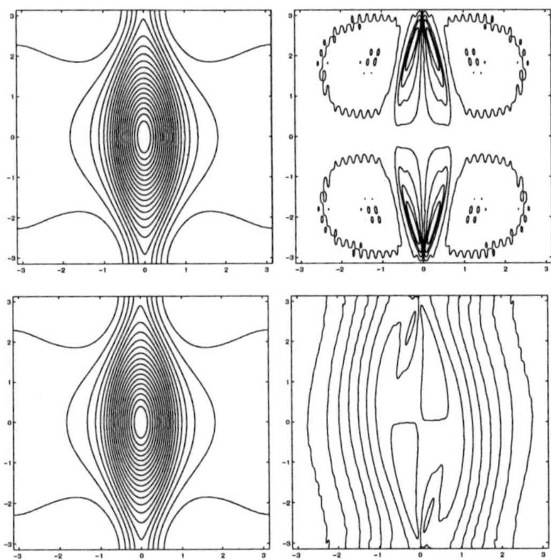

Fig. 7.   Contour plots of ψ and ϕ at $Pr = 0.2$, $\alpha = 2$ and $\epsilon = 2.44$ for P (up) and at $\epsilon = 2.435$ for Q equilibria (bottom).

When the pitchfork bifurcation for the equilibrium P takes place, the above conditions do not hold, therefore this symmetry is broken and the equilibria Q^* and

Q have only the weaker symmetry (13)-(14). Finally, we point out that for the equilibrium P the amplitudes related to the modes $(k_x, 0)$ and $(0, k_y)$ give a null contribute to the velocity field.

4. Conclusions and further developments

We have considered RRMHD for unreconnected equilibria with no motion

$$\psi_e^\alpha(x) = \frac{\psi_0^\alpha}{\cosh^2 \alpha x} \qquad x \in [-\pi, \pi], \quad \alpha \in R \tag{22}$$

Instability occurs by symmetry breaking bifurcation when $\epsilon < \alpha\sqrt{5}$, i.e. the stability threshold scales linearly with α. We have analyse the bifurcation diagram for $\alpha = 1$ and $\alpha = 2$.

For $\alpha = 1$, the bifurcation diagram and the topology of the fields are qualitatively the same obtained in the case of $\psi_e(x) = \cos x$.[22] Therefore the disappearance of small size islands because of a tangent bifurcation as a possible mechanism for "hard" excitation is confirmed.

For $\alpha = 2$ the equilibrium with a small island doesn't disappear by a tangent bifurcation, but it becomes unstable. The destabilization happens with a subcritical pitchfork bifurcation, involving a weakening in the symmetry of the fields. This is very relevant for the velocity field, which is not organized in convective cells, but becomes aligned with magnetic field. This configuration of the velocity field corresponds to higher kinetic energy and to stronger sensitivity to variations of the dissipation coefficients (viscosity and resistivity). The spontaneous development of zonal flows is considered relevant for fusion devices, since it can ameliorate transport. Indeed, sheared velocity layers are capable of suppressing turbulence induced by electrostatic drift instabilities, which seem to be responsible for anomalous transport in tokamak plasmas.

This new phenomenology is related to the increasing of the magnetic shear at $x = 0$. In order to explain the behavior of the magnetic and velocity fields varying the magnetic shear, further investigations are required. In particular, it is interesting to know the value of α for which the transition to higher kinetic energy equilibria takes place and to study the behavior of the system for much higher value of the magnetic shear. A remarkable difficulty consists in the computational complexity of the problem, which increases with the magnetic shear.

References

1. H.P. Furth, J. Kelleen and M.N Rosenbluth, *Phys. Flui ds*, **6**, 459, 1963.
2. D. Biskamp, *Nonlinear magnetohydrodynamics*, Cambridge University Press, Cambridge, 1993.
3. E. Priest and T. Forbes, *Magnetic Reconnection*, Cambridge University Press, 2000.
4. D. Grasso, L. Margheriti, C. Tebaldi, F. Porcelli, *Plasma Physics and Controlled Fusion*, **48**, L87 - L95, 2006.

5. See *e.g.* G. Falchetto and M. Ottaviani *Phys. Rev. Lett.* **92** (2) 025002-1, 2004 and other references quoted therein.

6. S.I. Braginskii, in *Rev. of Plasma Phys.* (edited by M.A. Leontovich), Vol.1, Consultants Bureau, New York, 1965.

7. R.Y. Dagazian and R.B. Paris, *Phys. Fluids* **29**, 1986.

8. H.P. Furth *et al*, *Phys. Fluids* **6**, 1963.

9. J. Guckenheimer and P. Holmes, *Nonlinear Oscillations, Dynamical Systems and Bifurcations of Vector Fields*, Springer, New York, 1986.

10. B.B. Kadomtsev and O.P. Pogutse, *Sov. Phys. JETP* **38**, 1974.

11. B.B. Kadomtsev, *Tokamak Plasma: a Complex System*, Institute of Physics Publishing, Bristol, 1992.

12. L. Margheriti and C. Tebaldi, in *Waves and Stability in Continuous Media*, World Sc., Singapore, 2002.

13. E.N. Parker, *J. Geophys. Res* **62**, 1957.

14. R.D. Parker *et al*, *Phys. Fluids* **B2**, 1990.

15. Y. Pomeau and P. Manneville, *Comm. Math. Phys.* **74**, 1980.

16. P.H. Rutherford, *Phys. Fluids* **16**, 1973.

17. A. Samain, *Plasma Phys. and Contr. Fusion* **26**, 1984.

18. B. Saramito and E.K. Maschke, in *Magnetic Turbulence and Transport* (P. Hennequin and M.A. Dubois eds.), Editions de Physique, Orsay, 1993.

19. P.A. Sweet, in *Electromagnetic Phenomena in Cosmic Physics* (edited by B. Lehnert), Cambridge University Press, Cambridge, 1958.

20. C. Tebaldi, in *Nonlinear Dynamics*, World Scientific, Singapore, 1989.

21. F. Porcelli, D. Borgogno, F. Califano, D. Grasso, M. Ottaviani and F. Pegoraro, *Plasma Phys. Control. Fusion*, **44**, B389, 2002.

22. C. Tebaldi *et al*, *Plasma Phys. and Contr. Fusion* **38**, 1996.

23. C. Tebaldi and M. Ottaviani, *Plasma Physics* **62**, 1999.

24. A. Thyagaraja, *Phys. Rev. Lett.* **24**, 1981.

25. F.L. Waelbroeck, *Phys. Rev. Lett.* **70**, 1993.

26. J.A. Wesson *et al*, in *Proc. of the Tenth Int. Conf. on Plasma Phys. and Controlled Nuclear Fusion Research*, Vol.2, IAEA, 1985.

27. J.A. Wesson *et al*, *Nucl. Fusion* **31**, 1991.

28. R.B. White *et al*, *Phys. Fluids* **20**, 1977.

ON A CLASS OF REACTION DIFFUSION SYSTEMS: EQUIVALENCE TRANSFORMATIONS AND SYMMETRIES

M. Torrisi* and R. Tracinà**

*Dipartimento di Matematica e Informatica, Università di Catania,
Italy*
E-mail: torrisi@dmi.unict.it
**E-mail: tracina@dmi.unict.it*

We get equivalence transformations for a class of reaction-diffusion systems. After having specialized them for a family of models concerned with bacterial colonies, we derive symmetries and some invariant solutions.

Keywords: Reaction-diffusion systems, equivalence transformations, bacterial colonies, symmetries

1. Introduction

We consider the quite general class of reaction-diffusion systems

$$u_t = (f(u,\, v)u_x)_x + g(u,v), \tag{1}$$

$$v_t = (h(u,v)v_x)_x + k(u,v), \tag{2}$$

where $u = u(t,x)$ and $v = v(t,x)$.

The reaction-diffusion systems are widely used in many applications of different disciplines. Examples, to name few, can be found in combustion problems, phase transitions, population dynamics, biological processes. In this paper motivated by the search for special solutions of a mathematical model, concerned with the development of bacterial colonies, we look for the equivalence transformations of aforesaid class of systems in order to get a symmetry classification of the bacterial colony model. Several classes of reaction-diffusion equations and systems have been considered in the framework of group analysis, a lot of references can be found in recent papers of A. G. Nikitin[1] and O.O. Vaneeva et al.[2]

As well known the group classification is a corner stone in the group analysis approach to the differential equations. The complete group classification gives the Lie symmetry algebras for all forms of the constitutive functions appearing in a model, but to get it is often not a simple task. The search for symmetries for a specific equation is usually a technical problem which can be solved trivially with the help of computer packages. This search becomes a wasteful venture when we

208

look for symmetries of equations where functions appear not *a priori* specified. That is when we look for the group classification of the class spanned from the arbitrary functions.

The *determining system* obtained by applying the Lie infinitesimal criterion is usually linear but, when the equations to classify include arbitrary functions, it becomes rather complicated and could show some nonlinearities.

A useful tool in performing a group classification of a class of equations is the knowledge of the equivalence transformation group.

An equivalence transformation for the system (1)-(2) is a non degenerate change of dependent and independent variables which brings into a new system having the same differential structure but where the functions f, g, h and k can have different form.

Of course the symmetries are special equivalence transformations.

The plan of the paper is the following. In the Section 2 we look for equivalence transformations of the system (1)-(2). In the Section 3 we show a reaction diffusion model for the development of bacterial colonies of *Proteus Mirabilis* and specialize the equivalence transformations for this model. Moreover, by using a suitable projection algorithm, we derive a symmetry classification and determine some classes of solutions for the model under consideration.

2. Equivalence algebra

An equivalence transformation is a non degenerate change of variables

$$t = t(\hat{t}, \hat{x}, \hat{u}, \hat{v}), \tag{3}$$

$$x = x(\hat{t}, \hat{x}, \hat{u}, \hat{v}), \tag{4}$$

$$u = u(\hat{t}, \hat{x}, \hat{u}, \hat{v}), \tag{5}$$

$$v = v(\hat{t}, \hat{x}, \hat{u}, \hat{v}), \tag{6}$$

bringing any system of the form (1)-(2) into a system of the same form but with different $\hat{f}(\hat{u}, \hat{v})$, $\hat{g}(\hat{u}, \hat{v})$, $\hat{k}(\hat{u}, \hat{v})$ and $\hat{h}(\hat{u}, \hat{v})$.

The obvious way to get equivalence transformations for the system (1)-(2) is the use of (3)-(6) in our system in order to get the form of the functions t, x, u, v. This way (see e.g. Winternitz and Gazeau[3,4] , Winternitz and Gagnon[5] and references inside) leads quite often to considerable difficulties and does not give always the general solutions.

A way to overcome these problems have been suggested by L.V.Ovsiannikov[6] which adopted a generalization of the Lie infinitesimal criterion to get the infinitesimal generator of the continuous equivalence transformations (see also the Ph.D. Thesis of I. G. Lisle[7]). In the following we will use the Ovsiannikov's generalization of the Lie criterion. The interested reader could see also additional information in[9–12] .

In the *basic augmented space* $A = X \times U \times P$, where $\{(t,x)\} = X \subseteq \mathcal{R}^2$, $\{(u,v)\} = U \subseteq \mathcal{R}^2$ and $\{(f,g,h,k)\} = P \subseteq \mathcal{R}^4$, let us consider the one parameter group of transformations

$$t = t(\hat{t}, \hat{x}, \hat{u}, \hat{v}, \varepsilon), \tag{7}$$

$$x = x(\hat{t}, \hat{x}, \hat{u}, \hat{v}, \varepsilon), \tag{8}$$

$$u = u(\hat{t}, \hat{x}, \hat{u}, \hat{v}, \varepsilon), \tag{9}$$

$$v = v(\hat{t}, \hat{x}, \hat{u}, \hat{v}, \varepsilon), \tag{10}$$

$$f = f(\hat{t}, \hat{x}, \hat{u}, \hat{v}, \hat{f}, \hat{g}, \hat{h}, \hat{k}, \varepsilon), \tag{11}$$

$$g = g(\hat{t}, \hat{x}, \hat{u}, \hat{v}, \hat{f}, \hat{g}, \hat{h}, \hat{k}, \varepsilon), \tag{12}$$

$$h = h(\hat{t}, \hat{x}, \hat{u}, \hat{v}, \hat{f}, \hat{g}, \hat{h}, \hat{k}, \varepsilon), \tag{13}$$

$$k = k(\hat{t}, \hat{x}, \hat{u}, \hat{v}, \hat{f}, \hat{g}, \hat{h}, \hat{k}, \varepsilon), \tag{14}$$

which is locally a C^∞-diffeomorphism, depends analytically on the parameter ε in a neighborhood of $\varepsilon = 0$ and reduces to the identity transformation for $\varepsilon = 0$.

The infinitesimal generator of the transformations (7)-(14) has the form

$$Y = \xi^1 \frac{\partial}{\partial t} + \xi^2 \frac{\partial}{\partial x} + \eta^1 \frac{\partial}{\partial u} + \eta^2 \frac{\partial}{\partial v} + \mu^1 \frac{\partial}{\partial f} + \mu^2 \frac{\partial}{\partial g} + \mu^3 \frac{\partial}{\partial h} + \mu^4 \frac{\partial}{\partial k} \tag{15}$$

where ξ^1, ξ^2, η^1 and η^2 are sought depending on t, x, u and v, while the μ^i ($i = 1, 2, 3, 4$) are sought depending on t, x, u, v, f, g, h and k.

By requiring the invariance of the system (1)-(2) with respect a suitable prolongation $Y^{(2)}$ of generator (15),

$$Y^{(2)}[u_t - (f(u, v)u_x)_x - g(u, v)] = 0, \tag{16}$$

$$Y^{(2)}[v_t - (h(u, v)v_x)_x - k(u, v)] = 0, \tag{17}$$

under the constraints that variables u and v have satisfy the system (1)-(2), following the well known techniques,[6–9,11] we get the *determining system* in the unknowns ξ^1, ξ^2, η^1, η^2 and μ^i ($i = 1, 2, 3, 4$).

At this step, in view of the further applications, we do not require the invariance of the auxiliary conditions[10,14,15] :

$$f_t = f_x = g_t = g_x = 0, \tag{18}$$

$$h_t = h_x = k_t = k_x = 0 \tag{19}$$

which specify the functional dependence of f, g, h and k. This implies that the obtained equivalence transformations change (1)-(2) into a system with the same differential structure but with the arbitrary functions f, g, h and k that could depend not only on u and v but also on t and x.

After having solved the *determining system* obtained from invariance conditions (16)-(17), we get the equivalence generator

$$Y = \phi(t)\frac{\partial}{\partial t} + (c_0 + xc_1)\frac{\partial}{\partial x} + (u\beta(t) + \gamma(t))\frac{\partial}{\partial u} + (v\psi(t) + \delta(t))\frac{\partial}{\partial v} +$$

$$+ (2c_1 f - f\phi_t)\frac{\partial}{\partial f} + (\gamma_t - g\phi_t + u\beta_t + g\beta)\frac{\partial}{\partial g} +$$

$$+ (\delta_t - h\phi_t + h\psi + v\psi_t)\frac{\partial}{\partial h} + (-2c_1 k - k\phi_t)\frac{\partial}{\partial k}, \tag{20}$$

where ϕ, ψ, β, γ and δ are arbitrary functions of their arguments.

The equivalence algebra $\mathcal{L}_\varepsilon$ is infinite-dimensional and is spanned by

$$Y_0 = \partial_x, \quad Y_1 = x\partial_x + 2f\partial_f - 2k\partial_k,$$
$$Y_\phi = \phi(t)\partial_t - f\phi_t\partial_f - g\phi_t\partial_g - k\phi_t\partial_k - h\phi_t\partial_h,$$
$$Y_\beta = u\beta(t)\partial_u + (u\beta_t + g\beta)\partial_g,$$
$$Y_\psi = v\psi(t)\partial_v + (h\psi + v\psi_t)\partial_h,$$
$$Y_\gamma = \gamma(t)\partial_u + \gamma_t\partial_g,$$
$$Y_\delta = \delta(t)\partial_v + \delta_t\partial_h.$$

3. On a model for development of Proteus Mirabilis bacterial colonies

Proteus mirabilis is a small *Gram-negative* bacterium part of the *Enterobacteriaceae* family. Commonly, it is part of the normal flora of the human intestinal tracts, but can also be found free living in water and soil. It has the power to shift his shape and often appears in different forms: *vegetative* cells and *swarmer* cells .

The organism goes through a cycle of *differentiation, migration* and *consolidation* depending upon the level of nutrients available to it.

When Proteus cells are inoculated on a surface of a suitable hard agar medium they grow as short vegetative rods.

These last ones, after their surface density v having reached a threshold value $v^\star$, start to differentiate at the colony margin into long *swarmer* cells possessing up to fifty times more flagella per unity of surface.

The swarmer cells, after a small time t_m, migrate rapidly away from the colony until they stop and revert by a series of cell fissions (*dedifferentiation*) into the vegetative cell form in a process called *consolidation*.

The waiting time t_m is spent from swarmer cells to aggregate in parallel arrays to form elongated hyper flagellated motile multicellular "rafts".

Only swarmer cells in contact with other swarmer cells are capable of translocation on the surface of a medium containing $\geq 1\%$ agar; vegetative cells and isolated swarmer cells are immobile.

Thus, swarm motility is an inherently cooperative process resulting in nonlinear transport of the population characterized by an expansion dependent on bacterial density.

Of course at the beginning time t_m, when the migration phase starts, the surface density u of swarmer cells has reached a value $u_m(> 0)$.

Based on experimental observations of cellular differentiation and group motility some models[16–21] has been developed to describe the swarmer cell differentiation-dedifferentiation cycle and the spatial evolution cycle, the spatial evolution of vegetative and swarmer cells during the *Proteus mirabilis* swarm colonies development.

Here we take in consideration the reaction diffusion model proposed by G.S. Medvedev, T.J. Kaper and N. Kopell.[19]

They used a method of modeling bacterial colonies development based on the reaction diffusion equation approach which describes the evolution of bacterial densities.

In general the rate of change of a bacterial density b can be described by:

$$\partial_t b = movement + "birth" - "death". \tag{21}$$

A canonical example of these equations is the well known Kolmogorov-Petroski-Piskounov-Fisher equation.

For the colony under consideration the motion is described by diffusion of the only swarmer cells, while the "birth" term corresponds to the bacterial reproduction (i.e. septation and differentiation) and "death" therm represents the transition of bacteria into a non moving state.

Then, after having denoted with u and v the surface densities of the swarmer and vegetative cells, respectively, we write the governing system, in 1+1 dimensions, as:

$$u_t = \nu v + (\alpha - \mu)u + (D(u,v)u_x)_x \tag{22}$$
$$v_t = (\alpha - \nu)v + \mu u, \tag{23}$$

where α, μ, ν are, in general, functions depending from the densities and characterizing respectively the cellular growth, septation and differentiation.

The diffusion appearing in the equation (22) models the migration of swarm cells. No diffusion appears, instead, in equation (23) due to the vegetative cells.

This model actually describes the behavior from the initial phase, where only vegetative cells appear, until the consolidation phase.

One of the problem of this model is to identify the different phases of the process by specializing the functional form of α, μ, ν, D.

Unfortunately analyzing the Proteus Mirabilis evolution using the aforesaid system appears to be far from trivial since one must solve an initial value problem for a system of two nonlinear equations.

Starting from this kind of model Medvedev, Kaper and Koppel[19] had studied by asymptotic methods the previous system by making the strong static approximation $v = v(x)$ and removing the second equation.

The goal of our study, instead, is to get information from the system not removing second equation. Our first approach to this system is the search for symmetries and then for invariant solutions. Even thought these latter ones could not solve the problem with its auxiliary conditions they can help in the characterization of the constitutive coefficients and can be useful as test solutions for a possible numerical approach.

We restrict our analysis to the migration phase.

In this phase α, μ and ν could be approximately assumed constants while D can be assumed depending of the density u and v.

In particular, when α, μ and ν are considered positive constants, we look for infinitesimal generator of a Lie group of point transformations for the system (22)-(23) by using the equivalence transformations obtained for the system (1)-(2).

In fact the system (22)-(23) can be written in the form (1)-(2) by putting

$$f(u, v) = D(u, v) \tag{24}$$
$$g(u, v) = \nu v + (\alpha - \mu)u \tag{25}$$
$$h(u, v) = 0 \tag{26}$$
$$k(u, v) = (\alpha - \nu)v + \mu u. \tag{27}$$

So in order to get equivalence algebras and symmetries for the system (22)-(23) we follow the procedures showed in[10,11,15] .

3.1. *Equivalence algebra of the model*

In the search for equivalence algebras for the system (22)-(23), we consider the projection $\tilde{Y}$ of the equivalence operator Y on the space (t, x, u, v, f). This projection is an equivalence operator for the system (22)-(23) if and only if the specializations of the functions g, h and k are invariant respect to Y.[15]

As a consequence we require the invariance of (25)-(27) with respect to the generator Y (15), that is

$$\mu^2 = \nu \eta^2 + (\alpha - \mu)\eta^1 \tag{28}$$
$$\mu^3 = 0 \tag{29}$$
$$\mu^4 = (\alpha - \nu)\eta^2 + \mu \eta^1, \tag{30}$$

under the constraints that the functions g, h and k satisfy (25)-(27).

By substituting in (28)-(30) the infinitesimal components η^1, η^2, μ^2, μ^3 and μ^4 of generator (20), we obtain further restrictions for the arbitrary functions ϕ, ψ, β, γ and δ appearing in the generator Y.

After solving these ones, we get that an equivalence operator for our system (22-23) is

$$\tilde{Y} = c_1 \partial_t + (c_3 x + c_0)\partial_x +$$
$$+ \left(c_1 u(\alpha - \nu) + c_2 u + c_4 \frac{\nu}{\mu} e^{\alpha t} - c_5 e^{(\alpha - \mu - \nu)t} \right) \partial_u + \tag{31}$$
$$+ \left(c_1 v(\alpha - \nu) + c_2 v + c_4 e^{\alpha t} + c_5 e^{(\alpha - \mu - \nu)t} \right) \partial_v + 2D c_3 \partial_D$$

where c_0, c_1, c_2, c_3, c_4 and c_5 are arbitrary constants.

3.2. *Symmetries*

In order to find the symmetry algebras for the system (22)-(23), we consider the projection X of the equivalence operator $\tilde{Y}$ (31) on the space (t, x, u, v). This projection is a symmetry operator for the system (22)-(23) if and only if the specialization of the function D is invariant respect to $\tilde{Y}$.[10-15] .

In this case we must require the invariance of

$$D = D(u, v) \tag{32}$$

with respect the generator $\tilde{Y}$, that is

$$\mu^1 = D_u \eta^1 + D_v \eta^2 \tag{33}$$

under the constraint that the function D satisfies (32).

Taking into account the infinitesimal components η^1, η^2 and μ^1 of the generator (31), the condition (33) becomes the following classifying equation for the function D

$$2c_3 D = D_u \left[c_1 u(\alpha - \nu) + c_2 u + c_4 \frac{\nu}{\mu} e^{\alpha t} - c_5 e^{(\alpha - \mu - \nu)t} \right] + \tag{34}$$

$$+ D_v \left[c_1 v(\alpha - \nu) + c_2 v + c_4 e^{\alpha t} + c_5 e^{(\alpha - \mu - \nu)t} \right].$$

So we obtain that the system (22)-(23) is invariant under a Lie group of point transformations with the following infinitesimal generator:

$$X = c_1 \partial_t + (c_3 x + c_0)\partial_x +$$

$$+ \left(c_1 u(\alpha - \nu) + c_2 u + c_4 \frac{\nu}{\mu} e^{\alpha t} - c_5 e^{(\alpha - \mu - \nu)t} \right) \partial_u +$$

$$+ \left(c_1 v(\alpha - \nu) + c_2 v + c_4 e^{\alpha t} + c_5 e^{(\alpha - \mu - \nu)t} \right) \partial_v$$

where c_0, c_1, c_2, c_3, c_4 and c_5 are related by the classifying equation (34) for the function D.

From (34) we get, easily, that the principal Lie Algebra $\mathcal{L_P}$ is two-dimensional and spanned by

$$X_1 = \partial_t, \qquad X_2 = \partial_x,$$

while the extensions with respect to $\mathcal{L_P}$ are the following:

(1) $D = \tilde{D}(\sigma)$ with $\tilde{D}$ arbitrary function of $\sigma = u + v$:

$$X_3 = e^{(\alpha - \mu - \nu)t} \partial_u - e^{(\alpha - \mu - \nu)t} \partial_v. \tag{35}$$

(2) $D = (u + v)^\gamma D_0$ where γ and D_0 are arbitrary constitutive constants: .

$$X_3 = e^{(\alpha - \mu - \nu)t} \partial_u - e^{(\alpha - \mu - \nu)t} \partial_v, \qquad X_4 = \frac{\gamma x}{2} \partial_x + u \partial_u + v \partial_v. \tag{36}$$

(3) $D = \tilde{D}(\sigma)$ with $\tilde{D}$ arbitrary function of $\sigma = \mu u - \nu v$:

$$X_3 = \nu e^{\alpha t} \partial_u + \mu e^{\alpha t} \partial_v. \tag{37}$$

214

(4) $D = (\sigma + D_1)^\gamma D_0$ where $\sigma = \mu u - \nu v$, while γ, D_0 and D_1 are arbitrary constitutive constants:

$$X_3 = \nu e^{\alpha t}\partial_u + \mu e^{\alpha t}\partial_v, \tag{38}$$

$$X_4 = \frac{\gamma x}{2}\partial_x + \left(u - \frac{D_1 e^{(\alpha-\mu-\nu)t}}{\mu+\nu}\right)\partial_u + \left(v + \frac{D_1 e^{(\alpha-\mu-\nu)t}}{\mu+\nu}\right)\partial_v. \tag{39}$$

(5) $D = (u - D_0)^\gamma \tilde{D}(\sigma)$ with $\tilde{D}$ arbitrary function of $\sigma = \frac{u-D_0}{v+D_0}$, while γ and D_0 are arbitrary constitutive constants:

$$X_3 = \frac{\gamma x}{2}\partial_x + \left(u - D_0 e^{(\alpha-\mu-\nu)t}\right)\partial_u + \left(v + D_0 e^{(\alpha-\mu-\nu)t}\right)\partial_v. \tag{40}$$

The diffusion coefficient

$$D = \hat{D}_0 \frac{u}{u + kv} \qquad (k > 0,\ \hat{D}_0 > 0), \tag{41}$$

suggested by Medvedev, Kaper and Koppel,[19] falls in this case, by putting $\gamma = D_0 = 0$. Moreover in this case, by putting $D_0 = 0$, $\gamma < 0$, $\tilde{D}_0 = const$, we find a diffusion coefficient of the form $D = \tilde{D}_0 u^\gamma$ which is a special case of $D = D(u^h)$ $(h < 0)$ suggested by A. Czirok, M. Matsushita and T. Vicsek.[18]

3.3. *Some invariant solutions*

For brevity reasons we consider here only the case

$$D = \frac{D_0}{u + v}$$

with D_0 positive constitutive constant.

This form of D belongs to the case 2 with $\gamma = -1$, so our system admits the following symmetry generators

$$X_1 = \partial_t, \quad X_2 = \partial_x,$$
$$X_3 = -x\partial_x + 2u\partial_u + 2v\partial_v, \quad X_4 = e^{(\alpha-\mu-\nu)t}(\partial_u - \partial_v).$$

The corresponding finite transformation is

$$\tilde{t} = \varepsilon_1 + t, \quad \tilde{x} = (\varepsilon_2 + x)e^{\varepsilon_3}, \tag{42}$$

$$\tilde{u} = u\,e^{2\varepsilon_3} - \varepsilon_4 e^{(\alpha-\mu-\nu)(\varepsilon_1+t)}, \quad \tilde{v} = v\,e^{2\varepsilon_3} + \varepsilon_4 e^{(\alpha-\mu-\nu)(\varepsilon_1+t)} \tag{43}$$

with ε_i $(i = 1 \ldots 4)$ arbitrary constants.

For this form of diffusion coefficient D it is a simple matter to find, after a trivial reduction, traveling wave solutions of the form

$$u = b_1 \frac{\nu}{\mu} e^{\alpha t - a_1 x}, \qquad v = b_1 e^{\alpha t - a_1 x} \tag{44}$$

with b_1, a_1 arbitrary constants and $b_1 > 0$.

If we apply the invariant finite transformation (42)- (43) to the previous solutions (44) one gets the new class of solutions

$$u = b_1 \frac{\nu}{\mu} e^{\alpha(t-\varepsilon_1)-a_1(xe^{\varepsilon_3}-\varepsilon_2)} e^{2\varepsilon_3} - \varepsilon_4 e^{(\alpha-\mu-\nu)t},$$

$$v = b_1 e^{\alpha(t-\varepsilon_1)-a_1(xe^{\varepsilon_3}-\varepsilon_2)} e^{2\varepsilon_3} + \varepsilon_4 e^{(\alpha-\mu-\nu)(t)},$$

which exhibits more arbitrary constants useful to satisfy the biological constraints of the model.

In this case we found also the following separate variables solutions

$$u = \frac{[b_1\nu\, e^{(\nu+\mu)t} - b_2\mu]\, e^{a_1 x+a_2}\, e^{(\alpha-\nu-\mu)t}}{\mu} \tag{45}$$

$$v = e^{a_1 x+a_2}\, [b_1\, e^{\alpha t} + b_2\, e^{(\alpha-\nu-\mu)t}] \tag{46}$$

with a_1, a_2, b_1, b_2 arbitrary constants.

The following additional solutions

$$u = \frac{[b_1\nu\, e^{(\nu+\mu)(t-\varepsilon_1)} - b_2\mu]\, e^{a_1(x\varepsilon_3-\varepsilon_2)+a_2}\, e^{(\alpha-\nu-\mu)(t-\varepsilon_1)}}{\mu} e^{2\varepsilon_3} - \varepsilon_4 e^{(\alpha-\mu-\nu)t}$$

$$v = e^{a_1(x\varepsilon_3-\varepsilon_2)+a_2}\, [b_1\, e^{\alpha(t-\varepsilon_1)} + b_2\, e^{(\alpha-\nu-\mu)(t-\varepsilon_1)}] e^{2\varepsilon_3} + \varepsilon_4 e^{(\alpha-\mu-\nu)t}$$

are also obtained by applying the same finite transformation (42)- (43) to the solutions (45)-(46).

4. Conclusions

We have taken in consideration a wide class of reaction-diffusion systems and by using a generalization of the Lie infinitesimal criterion we obtained its equivalence transformations. By restricting ourselves to a special subclass concerned with development of *Proteus Mirabilis* bacterial colonies, we derived an equivalence algebra for this last one which allows, by applying a projection algorithm, to get a symmetry classification with respect to the diffusion coefficient $D(u,v)$ of the swarmer cells. Finally some classes of invariant solutions are shown.

Acknowledgments

The Authors acknowledge the financial support from P.R.A. (ex 60%) of University of Catania, from G.N.F.M. of INdAM and from M.I.U.R. through the PRIN 2005/2007: *Nonlinear Propagation and Stability in Thermodinamycal Processes of Continuous Media*.

References

1. A.G. Nikitin, *Ukrainian Mathematical Bulletin* **2**, 153 (2005)
2. O.O. Vaneeva, A.G. Johnpillai, R.O. Popovych and C. Sophocleous, *Enhanced Group Analysis and Conservation Laws of Variable Coefficient Reaction-Diffusion Equations with Power Nonlinearities*, J. Math. Anal. Appl., in press, math-ph/0605081

3. P. Winternitz and J.P. Gazeau, *Phys. Lett. A*, **167**, 246, (1992).
4. J. P. Gazeau and P. Winternitz, *J. Math. Phys.*,**33**, 12, 4087 (1992).
5. L. Gagnon and P. Winternitz, *J. Phys. A*, **26**, 23, 7061 (1993).
6. L. V. Ovsiannikov, *Group Analysis of Differential Equations*, (New York: Academic, 1982).
7. I. G. Lisle, Equivalence transformations for classes of Differential Equations, Ph. D. Thesis, University of British Columbia (1992).
8. I. Sh. Akhatov, R. K. Gazizov and N. H. Ibragimov, *Mod. Probl. Math.*, **34**, 3 (in Russian) (1989).
9. N. H. Ibragimov, M. Torrisi and A. Valenti, *J. Math. Phys.*, **32**, 2988 (1991).
10. M. Torrisi, R. Tracinà and A. Valenti, *J. Math. Phys.*,**37**, 4758 (1996).
11. V. Romano and M. Torrisi, *J. Phys. A: Math. Gen.*, **32**, 45, 7953 (1999).
12. G. Gambino, A. M. Greco and M. C. Lombardo, *J. Phys. A: Math. Gen.*, **37**, 3835 (2004).
13. N. H. Ibragimov and M. Torrisi, *J. Math. Phys.*, **33**, 3931 (1992).
14. M. Torrisi and R. Tracinà, Equivalence transformations for Systems of First Order Quasilinear Partial Differential Equations, in Proc. Modern Group Analysis VI: Developments in Theory, Computation and Application (Edited by N. H. Ibragimov et F.M. Mahomed) (New Age International Publishers 1996).
15. M. Torrisi and R. Tracinà, *Int. J. Non-Linear Mechanics*, **33**, 473 (1998).
16. O. Rauprich, M. Matsushita, K. Weijer, F. Siegert, S. E. Esipov and J. A. Shapiro, *J. Bacteriol.*, **178**, 6525 (1996).
17. S.E. Esipov and J.A. Shapiro, *J.Math. Biol.*, **36**, 249 (1998).
18. A. Czirok, M. Matsushita and T. Vicsek, *Theory of periodic swarming of batteria: application to Proteus mirabilis,*,arXiv:physics/0007087 v1 28Jul 2000
19. G.S. Medvedev, T.J. Kaper and N. Kopell, *SIAM J. Appl. Math.* **60**, 5, 1601 (2000).
20. B. P. Ayati, *SIAM J. Numer. Anal.*, **37**, 1571 (2000).
21. B. P. Ayati, *J. Math. Biol.*, **52**, 1, 93 (2006).